Studies in Computational Intelligence

Volume 874

Series Editor

Janusz Kacprzyk, Polish Academy of Sciences, Warsaw, Poland

The series "Studies in Computational Intelligence" (SCI) publishes new developments and advances in the various areas of computational intelligence—quickly and with a high quality. The intent is to cover the theory, applications, and design methods of computational intelligence, as embedded in the fields of engineering, computer science, physics and life sciences, as well as the methodologies behind them. The series contains monographs, lecture notes and edited volumes in computational intelligence spanning the areas of neural networks, connectionist systems, genetic algorithms, evolutionary computation, artificial intelligence, cellular automata, self-organizing systems, soft computing, fuzzy systems, and hybrid intelligent systems. Of particular value to both the contributors and the readership are the short publication timeframe and the world-wide distribution, which enable both wide and rapid dissemination of research output.

The books of this series are submitted to indexing to Web of Science, EI-Compendex, DBLP, SCOPUS, Google Scholar and Springerlink.

More information about this series at http://www.springer.com/series/7092

Mohamed Abd Elaziz ·
Mohammed A. A. Al-qaness ·
Ahmed A. Ewees · Abdelghani Dahou
Editors

Recent Advances in NLP: The Case of Arabic Language

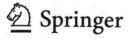 Springer

Editors
Mohamed Abd Elaziz
Faculty of Science
Mathematics Department
Zagazig University
Zagazig, Egypt

Ahmed A. Ewees
Computer Department
Damietta University
Damietta, Egypt

Mohammed A. A. Al-qaness
School of Computer Science
Wuhan University
Wuhan, China

Abdelghani Dahou
School of Computer Science
and Technology
Wuhan University of Technology
Wuhan, China

Mathematics and Computer
Science Department
Ahmed Draia University
Adrar, Algeria

ISSN 1860-949X   ISSN 1860-9503   (electronic)
Studies in Computational Intelligence
ISBN 978-3-030-34616-4   ISBN 978-3-030-34614-0   (eBook)
https://doi.org/10.1007/978-3-030-34614-0

This Springer imprint is published by the registered company Springer Nature Switzerland AG
The registered company address is: Gewerbestrasse 11, 6330 Cham, Switzerland

Foreword

The ongoing integration of various natural language processing (NLP) tasks including opinion mining, text summarization, machine translation, and document classification produces an important synergy of those tasks. Various solutions and applications as well as theoretical analyses and methodologies have been proposed for each task focusing on Arabic language. However, related problems to each research task and language open the opportunity for researchers to propose and develop new methods and systems to overcome them. This includes problems such as language complexity, language morphology, resource construction and availability, and language variation.

The main goal of this book is to provide the Arabic research community with the most advanced frameworks used in various Arabic NLP tasks. Moreover, this book presents several recent and state-of-the-art applications and methods related to Arabic NLP. A wide spectrum of research topics is presented in this book covering both the theoretical and the practical parts. Several fundamental aspects were discussed in this book including machine translation, speech recognition, morphological, syntactic, and semantic processing, information retrieval, text classification, text summarization, sentiment analysis, ontology construction, Arabizi translation, Arabic dialects, Arabic lemmatization, building and evaluating linguistic resources.

This book contains ten chapters selected for publication. These chapters have been reviewed based on several criteria and the reviewer's comments. The chapters are organized as follows: Chapter "Text Summarization: A Brief Review" presents a brief review concerning text summarization and the most recent studies related to Arabic text summarization. In this sense, chapter "Single Arabic Document Summarization Using Natural Language Processing Technique" presents single Arabic document summarization using NLP techniques. Meanwhile, chapter "A Proposed Natural Language Processing Preprocessing Procedures for Enhancing Arabic Text Summarization" provides a study on the impact of Arabic text preprocessing on text summarization. Chapter "Effects of Light Stemming on Feature Extraction and Selection for Arabic Documents Classification" presents an investigation on the effect of stemming techniques on feature extraction and selection in case of Arabic document classification. Chapter "Improving Arabic Lemmatization

Through a Lemmas Database and a Machine-Learning Technique" proposes a technique to improve Arabic lemmatization based on a lemmas database and a machine learning technique. In chapter "The Role of Transliteration in the Process of Arabizi Translation/Sentiment Analysis", the role of transliteration in the process of Arabizi translation and sentiment analysis has been studied. Chapter "Sentiment Analysis in Healthcare: A Brief Review" presents a brief review concerning sentiment analysis focusing on its applications in health care. Chapter "Aspect-Based Sentiment Analysis for Arabic Government Reviews" presents an Arabic aspect-based sentiment analysis that combines lexicon with rule-based models for Arabic Government Reviews. Chapter "Prediction of the Engagement Rate on Algerian Dialect Facebook Pages" studies the prediction of the engagement rate on Algerian dialect Facebook pages. In chapter "Predicting Quranic Audio Clips Reciters Using Classical Machine Learning Algorithms: A Comparative Study", a comparative study has been conducted on the prediction of Quranic audio clip reciters using machine learning.

This book is an important reference for researchers and practitioners interested in NLP and artificial intelligence, especially for Arabic language. This is due to the importance and the constant evolution of these fields. We would like to congratulate the authors for sending chapters, their effort, dedication, and contribution in this book. We would like to express our sincere gratitude to the reviewers for reviewing the chapters.

Hope this book will be a good material on the state-of-the-art research focusing on different Arabic NLP tasks.

Hubei, China   Songfeng Lu

Preface

With the emerging of new and advanced trends of natural language processing systems, various domains demand the employment and development of applications of these trends such as economics, education, and healthcare. During the process of developing these applications, many issues can be faced which are resolved by integrating language system solutions. For example, in online government services huge budgets are invested to develop systems that can automatically analyze clients' feedback and enhance their applications to meet clients' needs. Such kind of systems need the incorporation of artificial intelligence (AI) methods to automate their processes. The purpose of *Recent Advances in NLP: The Case of Arabic Language* edited book is providing the Arabic research community with the most advanced frameworks used in various Arabic NLP task optimizations while examining the theory and application of AI.

The edited book consists of a selection of an open call of chapters which target researchers, scholars, and students of the computational linguistics and natural language processing research community focusing on Arabic language. The aim of this book is intended to bring researchers from academia to present the state-of-the-art research activities concerning the recent advances, techniques, and applications of NLP and applied Arabic linguistics. Moreover, allowing the share of new ideas would guide current and future practitioners in the field of Arabic NLP.

We accepted ten chapters covering the following themes: machine translation, speech recognition, morphological, syntactic, and semantic processing, information retrieval, text classification, text summarization, sentiment analysis, ontology construction, Arabizi translation, Arabic dialects, Arabic lemmatization, building and evaluating linguistic resources. Several criteria were employed to assess and review each submission by the editorial board and a selected list of reviewers. These criteria include originality, contribution, technical quality, clarity of presentation, and interest and relevance to the book scope.

This book has been structured so that each chapter can be read independently from the others.

Chapter "Text Summarization: A Brief Review" presents a brief review concerning text summarization and the most recent studies related to Arabic text summarization.

Chapter "Single Arabic Document Summarization Using Natural Language Processing Technique" presents a single Arabic document summarization method using NLP techniques.

Chapter "A Proposed Natural Language Processing Preprocessing Procedures for Enhancing Arabic Text Summarization" provides a study on the impact of Arabic text preprocessing on text summarization. In particular, this chapter discusses Arabic preprocessing techniques including tokenization, normalization, stop-word removal, and structural processing.

Chapter "Effects of Light Stemming on Feature Extraction and Selection for Arabic Documents Classification" presents an investigation on the effect of stemming techniques on feature extraction and selection in case of Arabic document classification.

Chapter "Improving Arabic Lemmatization Through a Lemmas Database and a Machine-Learning Technique" reports on an attempt to construct an Arabic lemmatization system combining a lexicon-based approach with a machine learning-based approach. The goal is to implement a tool that does not need a morphological analyzer and returns all acceptable diacriticized for each word. Context-free lemmatization based on lexicon and contextual lemmatization using the hidden Markov model methods was adopted to implement the proposed SAFAR lemmatizer.

Chapter "The Role of Transliteration in the Process of Arabizi Translation/ Sentiment Analysis" reports on an attempt to study the impact of Arabizi transliteration on two tasks which are machine translation and sentiment classification. In particular, the authors tend to propose a rule-based Arabizi transliteration system as a preprocessing phase for Algerian dialect.

Chapter "Sentiment Analysis in Healthcare: A Brief Review" presents a brief review concerning sentiment analysis focusing on its applications in health care.

Chapter "Aspect-Based Sentiment Analysis for Arabic Government Reviews" presents an Arabic aspect-based sentiment analysis that combines lexicon with rule-based models to extract aspects of smart government applications Arabic reviews and classify them based on their corresponding sentiments.

Chapter "Prediction of the Engagement Rate on Algerian Dialect Facebook Pages" studies the prediction of the engagement rate on Algerian dialect Facebook pages using a proposed deep learning-based system. The system processes all the publication content: the text, the images, and the videos if they exist. In particular, two models of neural networks were proposed, one based on an MLP architecture and the other on a hybrid convolutional LSTM and MLP architecture.

Chapter "Predicting Quranic Audio Clips Reciters Using Classical Machine Learning Algorithms: A Comparative Study" presents a comparative study conducted on the prediction of Quranic audio clip reciters using machine learning. In particular, the study aims to evaluate and compare different classifiers performing the stated recognition including SVM, SVM linear, SVM-RBF, logistic regression, decision tree, random forest, ensemble AdaBoost, and extreme gradient boosting.

Finally, it is worth mentioning that this book serves a relatively small contribution to both NLP and Arabic language. Readers are encouraged to exploit and expand the information presented in order to improve it and apply it based on their needs.

Zagazig, Egypt   Mohamed Abd Elaziz
Damietta, Egypt   Ahmed A. Ewees
Wuhan, China   Mohammed A. A. Al-qaness
Wuhan, China   Abdelghani Dahou
August 2019

Contents

Abbreviations

ABSA	Aspect-based sentiment analysis
ACO	Ant colony optimization
AdaBoost	Ensemble Adaptive Boosting
ADS	Automatic document summarization
ALIF	Arabic Lexicon of Inflected Forms
ANLP	Arabic natural language processing
ATC	Automatic text classification
BoW	Bag of words
CA	Classical Arabic
CBOW	Continuous bag of words
Chi2	Chi-square
CNN	Convolutional neural network
DA	Dialect Arabic
DLs	Digital libraries
DT	Decision tree
EASC	Essex Arabic Summaries Corpus
FRAM	Frequency ratio accumulation method
GA	Genetic algorithm
HAAD	Human Annotated Arabic Dataset
HMM	Hidden Markov model
IE	Information extraction
IG	Information gain
IoT	Internet of things
IR	Information retrieval
ISRI	Information Science Research Institute
kNN	K-nearest neighbor
LDA	Linear discriminant analysis
LM	Language model
LR	Logistic regression
LSTM	Long short-term memory

MAE	Mean absolute error
MaxEnt	Maximum entropy
MBNB	Multi-variant Bernoulli Naive Bayes
MLE	Maximum likelihood estimation
MLP	Multilayer perceptron
MNB	Multinomial Naive Bayes models
MSA	Modern Standard Arabic
MSE	Mean squared error
NB	Naive Bayes
NLP	Natural language processing
OTE	Opinion target expression
POS	Part of speech
PV-DBOW	Distributed bag-of-words paragraph vectors
PV-DM	Distributed memory paragraph vectors
QDA	Quadratic discriminant analysis
RBF	Radial basis function
RF	Random forest
RNN	Recurrent neural network
ROUGE	Recall-Oriented Understudy for Gisting Evaluation
SA	Sentiment analysis
SALMA	Standard Arabic Language Morphological Analysis
SDG	Smart Dubai Government
SG	Skip-gram
SI	Swarm intelligence
SMO	Sequential Minimal Optimization
SMS	Short message system
SPA	Saudi Press Agency
SVD	Singular value decomposition
SVM	Support vector machine
SWEM	Smart Website Excellence Model
TC	Text classification
TF	Term frequency
tf–idf	Term frequency–inverse document frequency
UAE	United Arab Emirates
VAE	Variation autoencoder
VLEs	Virtual learning environments
XGBoost	Gradient boosting

Text Summarization: A Brief Review

**Laith Abualigah, Mohammad Qassem Bashabsheh, Hamzeh Alabool
and Mohammad Shehab**

Abstract Text Summarization is the process of creating a summary of a certain document that contains the most important information of the original one, the purpose of it is to get a summary of the main points of the document. Abstractive summarization of multi-documents aims to generate a concentrated version of the document while keeping the main information. Due to the massive amount of data these days, the importance of summarization arose. Finally, this paper collects the most recent and relevant research in the field of the text summarization to study and analysis for future research. It will be significant by giving a new direction to who are interested in this domain in the future.

Keywords Text summarization · Documents and processing · Main information · Arabic text summarization

1 Introduction

There is a huge amount of data surfacing digitally, therefore the importance of developing a punctuate procedure to shorten long texts immediately while keeping the main idea of it is necessary [1]. Summarization also helps shorten the time needed for reading, fasten the search for information and help to get the most amount of information on one topic [2, 3].

The central object of computerized text summarization is decreasing the reference text into a smaller version maintaining its knowledge alongside with its meaning. Several descriptions of text summarization are provided, for example [4] explained the report as text that is generated from one or more documents that communicate relevant knowledge in the first text, and that is no higher than half of the primary text(s) and usually significantly more limited than that.

L. Abualigah (✉) · M. Q. Bashabsheh · H. Alabool · M. Shehab
Faculty of Computer Sciences and Informatics,
Amman Arab University, Amman 11953, Jordan
e-mail: lx.89@yahoo.com

© Springer Nature Switzerland AG 2020
M. Abd Elaziz et al. (eds.), *Recent Advances in NLP: The Case of Arabic Language*, Studies in Computational Intelligence 874,
https://doi.org/10.1007/978-3-030-34614-0_1

As of late, content mining has turned into a fascinating exploration field because of the colossal measure of the existing content on the web [5]. Content mining is a basic field with regards to information digging for finding fascinating examples with regards to literary information. Inspecting and extricating of such data designs from enormous datasets is considered as an essential procedure [6, 7]. Many review studies were led to utilize different content digging strategies for unstructured datasets. It has been seen that the far-reaching overview considers in the Arabic setting were ignored. Belkebir and Guessoum [8] gave a wide survey of different examinations identified with Arabic content mining with more spotlight on the Holy Quran, assessment investigation, and web records. Moreover, the combination of the exploration issues and philosophies of the reviewed investigations will help the content mining researchers in seeking after their future examinations.

The study [9] determined text summarization as a small but true description of the contents of a text and according to [10], text summarization can be described as the manner of providing a shorter display of the most powerful information of a source or multiple references of information according to special demands. Text summarization techniques can be classified based on various models as explained in Fig. 1.

Text Summarization is the process of creating a summary of a certain document that contains the most important information of the original one, the purpose of it is to get a brief summary of the main points of the document [11]. Abstractive summarization of multi-documents aims to generate a concentrated version of the

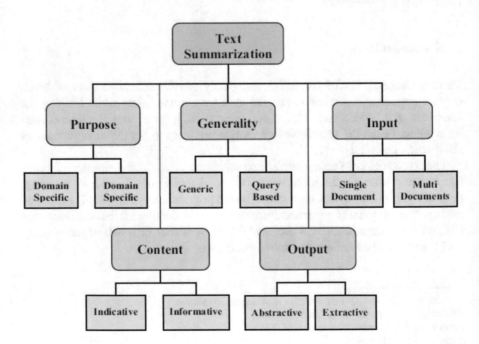

Fig. 1 Text summarization techniques [4]

Fig. 2 The number of
papers in the domain of text
summarization per year in
between (1999–2019), which
taken in this study

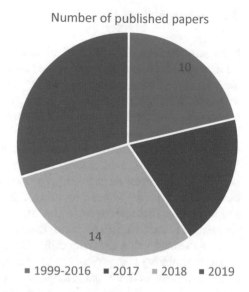

Number of published papers

■ 1999-2016 ■ 2017 ■ 2018 ■ 2019

document while keeping the main information [4]. Due to the massive amount of data these days, the importance of summarization arose [12].

Text summarization is a recent learning topic that caught attention rapidly, as research increase, we are hoping to witness a breakthrough that will affect this by providing a punctual method in summarizing long texts. In this paper, we provide a comprehensive review of text summarization techniques to draw attention to their importance in handling large data and to help researchers to use them to solve problems. Figure 2 shows the number of papers in the domain of text summarization per year in between (1999–2019), which taken in this study.

This paper is organized into three sections: Sect. 1 presented the introduction and organization of this paper for the text summarization domain. Section 2 Related Works shows the most related research in that domain. In Sect. 3 the related works are discussed and overviewed the used techniques. Finally, Sect. 3 concluded the paper to find the possible direction for the future researchers who are interested in the text summarization field.

2 Related Works

In this section, there are several papers in text summarization have been reviewed as following.

2.1   General Text Summarization

Because of the huge addition of information on the web, removing the most significant information as a reasonable brief would be profitable for specific clients. Accordingly, there is an enormous energy concerning the age of programmed content synopsis structures to establish abstracts consequently from the content, web, and interpersonal organization messages related with their satellite substance. This review features, just because, how the swarm intelligence (SI) advancement procedures are performed to settle the content summarization task proficiently [13–16].

Also, a persuading avocation regarding why SI, particularly Ant Colony Optimization (ACO), has been introduced [17]. Lamentably, three sorts of content synopsis undertakings utilizing SI show bit using in the writing when stood out from the other summarization strategies as AI and hereditary calculations, despite the way that there are genuinely encouraging results of the SI techniques. Then again, it has been seen that the summarization task with different kinds has not been formalized as a multi-target streamlining task previously, in spite of that there are numerous goals which can be considered. Additionally, the SI was not utilized before to help the continuous synopsis draws near. Subsequently, another model has been proposed to be satisfactory for accomplishing numerous targets and to fulfill the constant needs. In the long run, this investigation will enthuse analysts to further consider the different sorts of SI when tackling the synopsis assignments, especially, in the short content outline field.

There is a huge amount of data surfacing digitally, therefore the importance of developing a punctuate procedure to shorten long texts immediately while keeping the main idea of it is necessary [12]. Summarization also helps shorten the time needed for reading, fasten the search for information and help to get the most amount of information on one topic [18]. Text summarization is a fascinating learning topic that caught attention rapidly, as research increase, they are hoping to witness a breakthrough that will affect this by providing a punctual method in summarizing long texts [12].

The surge of information available through the internet and social networks and information technologies make the need of summarization more urgent, especially with the massive amount of data that is being spread due to the knowledge transfer among its users, which makes it difficult to differentiate between the right information from the wrong ones. To surmount the issues of data blast, a tool that can summarize these massive amounts of information has become a need. The procedure of summarization decreases the effort and time required to distinguish the most notable and important sentences. Generally, a summary can be described as content that is made from at least one message that passes on the most essential data in the first text while being adequately short [19].

Over the last two decades, the text summarization task has gained more significance due to the vast amount of online information, and its capability to extract helpful data and information in a manner that could be effectively taken care of people and utilized for a heap of purposes, including skilled systems for text appraisal.

This paper exhibits a programmed procedure for content evaluation that depends on fuzzy principles on an assortment of extracted information to locate the most critical data in the surveyed texts; the consequently delivered summaries of these texts are contrasted with reference summaries made by domain specialists. In contrast to different writing methods in the literature, this method summarizes manuscripts by researching correlated highlights to diminish dimensionality, and thus the number of fuzzy standards utilized for text summarization. Thus, the proposed methodology for content summarization with a moderately small number of fuzzy rules can profit advancement and utilization of future expert systems able to evaluate writing [20].

Strategies to automatically summarize, link and evaluate information have become more imperative as progressively substantial literary datasets have been gathered and made accessible by an assortment of (digital libraries (DLs), Virtual Learning Environments (VLEs), social media). Recently, Natural Language Processing (NLP) and Information Extraction (IE) techniques have been proposed to figure the comparability between free writings at low computational expenses [21].

Grammatical feature tagging is the procedure of consequently deciding the best possible linguistic tag or syntactic classification of a word contingent upon its specific situation. Part-of-Speech tagging is a fundamental advance in most Natural Language Processing (NLP) applications, for example, content summarization, question answering, data extraction and data recovery. A productive labeling approach for the Arabic language is proposed in [17] utilizing Bee Colony Optimization calculation. The issue is spoken to as a chart and a novel procedure is proposed to dole out scores to potential labels of a sentence, at that point the honey bees locate the best arrangement way. The proposed methodology is assessed utilizing KALIMAT corpus, which comprises of 18 M words. Trial results demonstrated that the proposed methodology accomplished 98.2% of exactness contrasted with 98, 97.4 and 94.6% for Hybrid, Hidden Markov Model and Rule-Based strategies separately. Moreover, the proposed methodology decided every one of the labels introduced in the corpus while the referenced methodologies can recognize just three labels. Other optimization techniques can be used [22–27].

Text summarization creates summaries from input documents while keeping striking data. It is an imperative task and can be connected to a few true applications. Numerous strategies have been proposed to take care of the content summarization issue [28, 29]. There are two principles for the text summarization systems: extractive and abstractive. Extractive summarization creates synopsis by choosing remarkable sentences or expressions from the source content, while abstractive strategies reword and rebuild sentences to form the summary. They focus around abstractive summarization in this work as it is increasingly adaptable and hence can create progressively different summaries [30].

Recent neural system ways to deal with an outline are generally either sentence-extractive, choosing a lot of sentences as the summary, or abstractive, creating the summary from a seq 2seq model. In this work [31], they present a neural model for single-record summary dependent on joint extraction and pressure. Following later fruitful extractive models, they outline the summarization issues as a progression of local choices. This model picks sentences from the report and after that chooses which of a set of compression options to apply to each chosen sentence. They compute this arrangement of discrete compression rules dependent on syntactic constituency parses; however, the proposed methodology is measured and it could utilize for any accessible source of compressions. For learning, they build oracle extractive-compressive summaries that reflect vulnerability over the proposed model's decision sequence, and then learn both of these parts together with this supervision. Test results on the CNN/Daily Mail and New York Times datasets demonstrate that this model accomplishes the innovative execution on substance determination assessed by ROUGE. Besides, human and manual assessment demonstrate that the proposed model's yield, for the most part, remains.

Recent days, a large portion of the client inquiries are of the complex in nature because of an increased need for dynamic information [32]. The overwhelming questions are replied with a summarization of applicable sentences from the web archives. These can be responded to by network question answering (CQA, for example, Quora, stack flood, Yahoo! Answers and so forth for different client levels utilizing AI methods. Answers can be created from different sources and summarized from a list of sentences [33] proposes a novel summarization technique which centers around programmed machine-produced summaries. Content summaries are the method utilized for condensing content with special, pertinent highlights in compression ranges.

The learning model is prepared with benchmark datasets 20 newsgroup and DUC2001 using machine learning algorithms. The examinations are completed and checked with standard measurements, for example, ROUGE for the results [34]. Primary categorization of text summarization techniques is based on the type of summary generated. It can either be of extractive or abstractive type. Generating abstractive summary is cumbersome as it gives a summary with sentences different from the original document, though the meaning of information is preserved. On the other hand, Extractive text summarization uses sentences from the document to provide a condensed form of the document that is in simple terms, it is the subset of the actual document. Most of the studies on text summarization are on extractive techniques [35].

Because of countless records accessible on the web, messages and advanced libraries, archive grouping is turning into a significant assignment incredibly required. It is ordinarily accomplished in the wake of performing highlight choice that comprises of choosing proper highlights to improve the grouping exactness. The vast majority of the component determination put together content order strategies depend with respect to building a term-frequency converse record frequency include vector, which is not typically proficient. Likewise, various archive order studies are centered on the English language. Garg and Saini [36] proposed an Arabic Text Classification technique, which is not seriously examined because of the multifaceted nature of the Arabic language. Another firefly calculation-based component choice strategy is proposed. This calculation has been effectively connected to various combinatorial issues. In any case, it has not been associated with highlight choice idea to manage Arabic Text Classification. To approve this method, Support Vector Machine classifier is utilized just as three assessment measures including exactness, review, and F-measure. Besides, probes OSAC genuine dataset alongside an examination with the innovative strategies are performed. The proposed strategy accomplishes an exactness worth equivalent to 0.994. The outcomes affirm the effectiveness of the proposed highlight choice technique in improving Arabic Text Classification precision.

The objective of the individual in summarizing any text is to get objective and concise ideas around it. It depending on the audience that wrote the text for it, in addition to the size of the original text, and there are many reasons that urge the individual to summarize a text: Summary for study, where the individual works to write important points in the text in order to learn the required material quickly. Summarizing the text helps readers read and understand it. The abstract is also used to write papers and academic papers to review and summarize the information written about the subject. There are grounds to consider when summarizing any text: Mention all important concepts and information in the text, and do not mention any non-important information. Do not repeat any information, delete all duplicate information. Replace difficult terms with simpler terms. Select and write the main idea in the text. Examples of summarizing the text (summarizing the text of the Indian language and the text on foreign languages [37].

2.2 Arabic Text Summarization

Arabic text clustering is an essential job for getting high-grade results with the common Information Retrieval (IR) systems particularly with the fast growth of the number of documents existing in the Arabic language. Documents clustering tries to automatically group related documents in one cluster working various similarity/distance criteria [38]. The length of the document often influences this job, useful knowledge on the documents is often followed by a large amount of noise, and since it is important to reduce this noise while having useful knowledge to increase the achievement of text documents clustering.

In this paper [39], they suggested assessing the impact of text summarization utilizing the Latent Semantic Analysis Model in order to determine problems indicated above, using 5 similarity/distance criteria without and with stemming. The experimental results show that the proposed method completely explains the problems of noisy knowledge and documents time, and thus significantly enhance clustering production.

Marie-Sainte and Alalyani [1] proposed the utilization of Firefly calculation for the extraction of summarization of single Arabic records. The proposed methodology is contrasted and two transformative methodologies that utilization hereditary calculations and concordance search. The EASC Corpus and the ROUGE toolbox are utilized for the assessment of the proposed methodology. Trial results demonstrated that the proposed methodology accomplished focused and significantly higher ROUGE scores in correlation with the two innovative draws near.

Content synopsis is the way toward creating a shorter variant of particular content. Programmed synopsis methods have been connected to different spaces, for example, therapeutic, political, news, and legitimate areas demonstrating that adjusting area applicable highlights could improve the outline execution. Regardless of the presence of a lot of research work in the area-based synopsis in English and different dialects, such work is absent in Arabic because of the lack of existing information bases.

In this paper [4], a half-breed, single-report content summarization approach is introduced. The methodology consolidates area information, factual highlights, and hereditary calculations to extricate significant purposes of Arabic political records. The ASDKGA approach is tried on two corpora KALIMAT corpus and Essex Arabic Summaries Corpus (EASC). The Recall-Oriented Understudy for Gisting Evaluation (ROUGE) structure was utilized to analyze the naturally produced synopses by the ASDKGA approach with outlines created by people. Likewise, the methodology is thought about against three other Arabic content outline draws near. This proposed approach showed promising outcomes when abridging Arabic political records with a normal F-proportion of 0.605 at the pressure proportion of 40%.

The exponential development of online literary information set off the critical requirement for a compelling and useful asset that consequently gives the ideal substance in an abridged structure while safeguarding center data.

In this paper [5], a programmed, nonexclusive, and extractive Arabic single archive outlining strategy is proposed, which goes for delivering an adequately enlightening summarization. The proposed extractive strategy assesses each sentence dependent on a mix of measurable and semantic highlights in which a novel definition is utilized considering sentence significance, inclusion, and assorted variety. Further, two abridging procedures including score-based and managed AI were utilized to create the summarization and afterward help to utilize the structured highlights. They exhibit the viability of the proposed strategy through a lot of investigations under EASC corpus utilizing ROUGE measure. Contrasted with some current related work, the test assessment demonstrates the quality of the proposed technique as far as exactness, review, and F-score execution measurements.

Arabic Text report bunching is a significant perspective for giving theoretical route and perusing systems by arranging monstrous measures of information into few characterized groups. In any case, Words as the vector are utilized for grouping techniques is regularly inadmissible as it overlooks connections between significant terms. Bunch examination isolates information into gatherings on groups for improved comprehension or summarization. Grouping has a long history and numerous strategies created in measurements, information mining, design acknowledgment and different fields. This examination proposes three approaches; Unsupervised, Semi Supervised procedures and Semi Supervised with dimensionality decrease to develop a grouping-based classifier for Arabic content archives.

Utilizing k-means, steady k-means Threshold + k-means and k- means with dimensionality decrease, after report preprocessing evacuating stop words and gets the root for each term in each record [40]. At that point, apply a term weighting technique to get the heaviness of each term concerning its record. At that point, apply a similitude measure technique to each report and its likeness with different archives. In addition, utilizing F-measure, entropy and bolster vector machine for figure precision. The datasets are online dynamic datasets that are portrayed by its accessibility and believability on the web. Arabic language is a difficult dialect when connected in a deduction-based calculation. In this way, choosing the fitting dataset is a chief factor in such research. The precision of those techniques contrasted and different methodologies and the proposed strategies shows better exactness and less blunders for new arrangement experiments. Taking into account that the measurement decrease procedure is exceptionally delicate because expanding the proportion of decrease can crush significant terms.

Text summarization is one of the most testing and troublesome undertakings in normal language handling, and computerized reasoning even more largely. Different methodologies have been proposed in the writing. Content synopsis is ordered into two classes: extractive content outline and abstractive content summarization. Most by far of work in the writing pursued the extractive methodology, likely because of the intricacy of the abstractive one. As far as we could know, the work displayed here is the main work on Arabic that handles both the extractive and abstractive angles. To be sure, while the writing needs synopsis structures that permit the mix of different tasks inside a similar framework. It additionally gives a component that permits the task of the appropriate activity to each part of the source content, which is to be condensed, and this is accomplished in an iterative procedure.

Mosa et al. [12] examined an element based synopsis for digging sentiment for Arabic audits so as to produce an outline that co contains a lot of positive surveys to a specific element, just as negative audits for a similar component. This examination fundamentally relies upon the Natural Language Processing, beginning with extricating an element for a particular area, at that point estimation arrangement and a while later it condenses these audits as per the highlights.

Conventional Arabic content outline frameworks depend on pack of-words portrayal, which includes meager and high-dimensional information. In this manner, dimensionality decrease is enormously expected to expand the intensity of highlights separation. In this paper, they present another strategy for ATS utilizing variation auto-encoder (VAE) model to take in an element space from high-dimensional information.

Alami et al. [40] investigated a few information portrayals, for example, term frequency (tf), tf-idf and both neighborhood and worldwide vocabularies. All sentences are positioned by the inert portrayal created by the VAE. They examine the effect of utilizing VAE with two summarization draws near, chart-based, and question-based methodologies. Analyses on two benchmark datasets explicitly intended for ATS demonstrate that the VAE utilizing tf-idf portrayal of worldwide vocabularies gives an increasingly discriminative component space and improves the review of different models. Test results affirm that the proposed technique prompts preferred execution over a large portion of the best in class extractive synopsis approaches for both charts based and inquiry-based outline draws near.

Computerized summarization help handle the consistently developing volume of data coasting around. There are two general classes: concentrate and conceptual. In the previous, they hold the more significant sentences pretty much in their unique structure, while the last requires a combination of numerous sentences as well as summarizing. This is a more testing errand than concentrate summarization.

In this paper [41], a novel conventional theoretical summarizer is displayed for a solitary report in the Arabic language. The framework begins by dividing the info content subject shrewd. At that point, each literary section is extractive abridged. Finally, they apply rule-based sentence decrease procedure. The RST-based extractive summarizer is an improved variant of the framework. By controlling the size of the extricated outline of each fragment, they can top the size of the last abstractive synopsis. Both summarizers, the improved extractive and the abstractive, were assessed. They tried to improve extractive summarizer on the equivalent dataset in the previously mentioned paper, utilizing the measures review, accuracy, and Rouge. The outcomes show detectable improvement in the exhibition, particularly the accuracy in shorter summarization. The abstractive summarizer was tried on a lot of 150 reports, creating synopses of sizes half, 40, 30 and 20% (of the first's statement check). Two human specialists who reviewed them out of the extreme score of five evaluated the outcomes. The normal score extended somewhere in the range of 4.53 and 1.92 for synopses at various granularities, with shorter outlines accepting the lower score. The trial results are empowering and exhibit the viability of the proposed methodology.

Supposition mining applications work with countless conclusion holders. This implies a summarization of assessments is significant so we can without much of a stretch translate holders' sentiments. This paper expects to give a component-based synopsis to Arabic audits. In this work [42], a framework is proposed utilizing Natural Language Processing (NLP) procedures, data extraction, and slant vocabularies. This gives clients to get to the feelings communicated in many surveys in a brief and valuable way. They begin with extricating highlight for a particular space, appointed opinion characterization to each component, and afterward abridged the audits. They led many trials to assess the proposed framework by utilizing information corpus from the lodging area. The exactness for conclusion mining they determined utilizing target assessment was 71.22%. They, likewise, connected emotional assessment for the synopsis age and it showed that the proposed framework accomplished an important proportion of 73.23% precision for positive outline and 72.46% exactness for a negative summarization.

3 Discussions and Overview

There are many challenges facing text summarization. Table 1 shows some details of the text summarization studies that conducted to handle the problem of text extraction and abstraction. Notice that the majority of reviewed studies focused on the problem of text extraction over abstraction. However, text abstraction is more difficult than extraction since it refers to the reformulated version of the original text. While text extraction studies aim to extract a copy of some sentences of the original text. In addition, the majority of the reviewed studies are proposed a singular method (e.g., Fuzzy-based method, Score-based method, Cluster-based method, Semantic-based method, Optimization-Based method, and Machine learning-based method) over proposing hybrid method.

Table 1 studies on methods for text summarization

References	Year	Output	Method	Language	Input Type	Validation	Compared with	Evaluation
[20]	2019	Extractive	Fuzzy method	Brazilian Portuguese texts	Multi documents	Experiment	Naive baseline, score, model and sentence	ROUGE
[38]	2012	Abstractive	Fuzzy method	English	Multi documents	Example	N/A	N/A
[32]	2018	Abstractive	Neural network-based methods/bottom-up attention	English	Single document	Experiment	Graph based attention neural model and	ROUGE
[19]	2013	Abstractive	Latent Semantic Analysis	Arabic	Single document	Experiment	N/A	N/A
[4]	2018	Extractive	Hybrid approach (domain knowledge, statistical features, and genetic algorithms)	Arabic	Single document	Experiment	Graph-based approach, statistical-based approach), and lexical cohesion and text entailment	Human and ROUGE
[34]	2019	Extractive	Hybrid approach (score-based method and machine-learning method.)	Arabic	Single document	Experiment	Score-based, cluster-based, semantic-based, optimization-based and machine learning based	ROUGE
[5]	2017	Extractive	β-Hill climbing search	English	Single document	Experiment	Original hill climbing search	P, R and F1

(continued)

Table 1 (continued)

References	Year	Output	Method	Language	Input Type	Validation	Compared with	Evaluation
[42]	2017	Abstractive	Rule-based sentence reduction technique	Arabic	Single document	Experiment	Ikhtasir approach	Human and ROUGE
[12]	2018	Extractive	Machine learning based	Arabic	Single document	Experiment	N/A	Human and P, R and F1
[10]	2019	Extractive	Firefly algorithm based feature selection method	Arabic	Single document	Experiment	InfoGain, OneR and TFIDF.	P, R and F1
[40]	2018	Extractive	Variation auto-ncoder (VAE) model	Arabic	Single document	Experiment	Latent semantic analysis (LSA), LexRank, Graph-based AE, TextRank, baseline Tf.ISF	Human and ROUGE

The hybrid method might achieve better results in terms of recall, precision, and F-measure than using one single method for handling text summarization. This is because the process of text summarization is a multi-dimensional problem that combines: segmentation and tokenization, word and sentence scoring, text evaluation, Cosine Similarity Matrix, identify the optimal sentence combination, which needs to be handled by different methods. Furthermore, most of the proposed methods are used to extract text from single documentation over multi documentation that could provide important information about the original text. Finally, two evaluation methods are used: human-based and automatic based (e.g., ROUGE). Studies should tack human evaluation in consideration since this approach is that it assesses coherence and informatively of the summary compared to the original text.

4 Conclusions

Due to the tremendous increment of data on the web, extracting the most important data as a conceptual brief would be valuable for certain users. Because of the large volume of data deployed in digital space, there is a need to find a way to shorten texts and provide clear summaries. Summarizing the texts is still active in several research and needs further research and development in summarizing the texts Due to the huge increase in data on the web, extracting the most important data as a conceptual summary will be useful to some researchers. In this paper, we drew the reader to the latest and most important data problems and the need to summarize the texts and explained the utility of short texts while preserving the original texts. This paper provides a summary of future research findings in this area. Finally, the optimization way can be used to deal with different problems.

References

1. S.L. Marie-Sainte, N. Alalyani, Firefly algorithm-based feature selection for Arabic text classification. J. King Saud Univ.-Comput. Inf. Sci. (2018)
2. L.M. Abualigah, A.T. Khader, M.A. Al-Betar, Multi-objectives-based text clustering technique using K-mean algorithm, in *2016 7th International Conference on Computer Science and Information Technology (CSIT)* (IEEE, 2016), pp. 1–6
3. A.M. Azmi, N.I. Altmami, An abstractive Arabic text summarizer with user-controlled granularity. Inf. Process. Manage. **54**(6), 903–921 (2018)
4. D.R. Radev, E. Hovy, K. McKeown, Introduction to the special issue on summarization. Comput. Linguist. **28**(4), 399–408 (2002)
5. A. Qaroush, I.A. Farha, W. Ghanem, M. Washaha, E. Maali, An efficient single document Arabic text summarization using a combination of statistical and semantic features. J. King Saud Univ.-Comput. Inf. Sci. (2019)
6. J. Xu, G. Durrett, Neural extractive text summarization with syntactic compression. arXiv preprint arXiv:1902.00863 (2019)
7. L.M.Q. Abualigah, Feature selection and enhanced krill herd algorithm for text document clustering. studies in computational intelligence (2019)

8. R. Belkebir, A. Guessoum, TALAA-ATSF: a global operation-based arabic text summarization framework, *Intelligent Natural Language Processing: Trends and Applications* (Springer, Cham, 2018), pp. 435–459
9. H. Saggion, Automatic summarization: an overview. Revue française de linguistique appliquée **13**(1), 63–81 (2008)
10. R.Z. Al-Abdallah, A.T. Al-Taani, Arabic text summarization using firefly algorithm, in *2019 Amity International Conference on Artificial Intelligence (AICAI)*. (IEEE, 2019), pp. 61–65
11. X. Carreras, L. Màrquez, Introduction to the CoNLL-2004 shared task: Semantic role labeling, in *Proceedings of the Eighth Conference on Computational Natural Language Learning (CoNLL-2004) at HLT-NAACL 2004* (2004), pp. 89–97
12. M.A. Mosa, A.S. Anwar, A. Hamouda, A survey of multiple types of text summarization with their satellite contents based on swarm intelligence optimization algorithms. Knowl.-Based Syst. **163**, 518–532 (2019)
13. L.M. Abualigah, A.T. Khader, E.S. Hanandeh, Hybrid clustering analysis using improved krill herd algorithm. Appl. Intell. (2018)
14. L.M. Abualigah, A.T. Khader, M.A. Al-Betar, O.A. Alomari, Text feature selection with a robust weight scheme and dynamic dimension reduction to text document clustering. Expert Syst. Appl. **84**, 24–36 (2017)
15. L.M. Abualigah, A.T. Khader, M.A. AlBetar, E.S. Hanandeh, Unsupervised text feature selection technique based on particle swarm optimization algorithm for improving the text clustering, in *Eai International Conference on Computer Science and Engineering* (2017)
16. L.M. Abualigah, A.T. Khader, M.A. Al-Betar, Z.A.A. Alyasseri, O.A. Alomari, E.S. Hanandeh, Feature selection with β-hill climbing search for text clustering application, in *2017 Palestinian International Conference on Information and Communication Technology (PICICT)* (IEEE, 2017), pp. 22–27
17. A. Alhasan, A.T. Al-Taani, POS tagging for arabic text using bee colony algorithm. Procedia Comput. Sci. **142**, 158–165 (2018)
18. R. Barzilay, M. Elhadad, Using lexical chains for text summarization. Adv. Autom. Text Summ. 111–121 (1999)
19. A. Ibrahim, T. Elghazaly, Improve the automatic summarization of Arabic text depending on Rhetorical Structure Theory, in *2013 12th Mexican International Conference on Artificial Intelligence* (IEEE, 2013), pp. 223–227
20. F.B. Goularte, S.M. Nassar, R. Fileto, H. Saggion, A text summarization method based on fuzzy rules and applicable to automated assessment. Expert Syst. Appl. **115**, 264–275 (2019)
21. R. Nallapati, F. Zhai, B. Zhou, Summarunner: a recurrent neural network based sequence model for extractive summarization of documents, in *Thirty-First AAAI Conference on Artificial Intelligence* (2017)
22. L.M.Q. Abualigah, E.S. Hanandeh, Applying genetic algorithms to information retrieval using vector space model. Int. J. Comput. Sci., Eng. Appl. **5**(1), 19 (2015)
23. L.M. Abualigah, A.T. Khader, Unsupervised text feature selection technique based on hybrid particle swarm optimization algorithm with genetic operators for the text clustering. J. Supercomput. **73**(11), 4773–4795 (2017)
24. L.M. Abualigah, A.T. Khader, E.S. Hanandeh, A new feature selection method to improve the document clustering using particle swarm optimization algorithm. J. Comput. Sci. (2017)
25. L.M. Abualigah, A.T. Khader, E.S. Hanandeh, A combination of objective functions and hybrid krill herd algorithm for text document clustering analysis. Eng. Appl. Artif. Intell. (2018)
26. L.M. Abualigah, A.T. Khader, E.S. Hanandeh, A novel weighting scheme applied to improve the text document clustering techniques, *Innov. Comput., Optim. Its Appl.* (Springer, Cham, 2018), pp. 305–320
27. L.M. Abualigah, A.T. Khader, E.S. Hanandeh, A.H. Gandomi, A novel hybridization strategy for krill herd algorithm applied to clustering techniques. Appl. Soft Comput. **60**, 423–435 (2017)

28. B. Ionescu, H. Müller, M. Villegas, A.G.S. de Herrera, C. Eickhoff, V. Andrearczyk, … Y. Ling, Overview of Image CLEF 2018: challenges, datasets and evaluation, in *International Conference of the Cross-Language Evaluation Forum for European Languages* (Springer, Cham, 2018), pp. 309–334
29. Z.A. Al-Sai, L.M. Abualigah, Big data and E-government: a review, in *2017 8th International Conference on Information Technology (ICIT)* (IEEE, 2017), pp. 580–587
30. H. Zhang, Y. Gong, Y. Yan, N. Duan, J. Xu, J. Wang, … M. Zhou (2019) Pretraining-based natural language generation for text summarization. arXiv preprint arXiv:1902.09243
31. K. Karpagam, A. Saradha, A framework for intelligent question answering system using semantic context-specific document clustering and Wordnet. Sādhanā **44**(3), 62 (2019)
32. Y. Wu, Pattern-enhanced topic models and relevance models for multi-document summarisation (Doctoral dissertation, Queensland University of Technology, 2018)
33. H. Froud, A. Lachkar, S.A. Ouatik, Arabic text summarization based on latent semantic analysis to enhance arabic documents clustering. arXiv preprint arXiv:1302.1612 (2013)
34. A. Kumar, A. Sharma, Systematic literature review of fuzzy logic based text summarization. Iran. J. Fuzzy Syst. 1706–3727 (2019)
35. A. Garg, J.R.A. Saini, Systematic and exhaustive review of automatic abstrac-tive text summarization for hindi language (2012)
36. A.K. Sangaiah, A.E. Fakhry, M. Abdel-Basset, I. El-henawy, Arabic text clustering using improved clustering algorithms with dimensionality reduction. Clust. Comput. 1–15 (2018)
37. S. Shashikanth, S. Sanghavi, Text summarization techniques survey on telugu and foreign languages (2019)
38. A. Wilbik, J.M. Keller, A fuzzy measure similarity between sets of linguistic summaries. IEEE Trans. Fuzzy Syst. **21**(1), 183–189 (2012)
39. Q.A. Al-Radaideh, D.Q. Bataineh, A hybrid approach for Arabic text summarization using domain knowledge and genetic algorithms. Cogn. Comput. **10**(4), 651–669 (2018)
40. N. Alami, N. En-nahnahi, S.A. Ouatik, M. Meknassi, Using unsupervised deep learning for automatic summarization of Arabic documents. Arab. J. Sci. Eng. **43**(12), 7803–7815 (2018)
41. A.M. El-Halees, D. Salah, Feature-based opinion summarization for arabic reviews, in *2018 International Arab Conference on Information Technology (ACIT)*. (IEEE, 2018), pp. 1–5
42. D.I. Saleh, feature-based opinion summarization for Arabic reviews. Featur.-Based Opin. Summ. Arab. Rev. (2017)

Single Arabic Document Summarization Using Natural Language Processing Technique

Asmaa A. Bialy, Marwa A. Gaheen, R. M. ElEraky, A. F. ElGamal and Ahmed A. Ewees

Abstract This paper presents a method based on natural language processing (NLP) for single Arabic document summarization. The suggested method based on the extractive method to select the most valuable information in the document. However, working with Arabic text is considered as a challenging task, this chapter tries to produce an accurate result by using some of NLP techniques. The proposed method is formed from three phases, the first one work as a pre-processing phase to unify synonyms terms, stemming, remove punctuation marks and remove text decoration. Consequently, it produces the features vectors and scores these features to start to select the clauses with the highest scores then marks it as important clauses. The suggested method's results are compared versus the traditional methods. In this context, two human experts summarized all the datasets manually in order to prepare a strong compare and effective evaluation of the suggested method. In the evaluation phase, some of the performance measures include accuracy, precision, recall, f-measure, and Rouge measure are used. The experimental results denoted that the suggested method showed a competitive execution compared with the human experts in summarization ratio as well as in the accuracy of the produced document.

Keywords Natural language processing · Arabic text summarization · Single document summarization · Extractive method

A. A. Bialy · M. A. Gaheen · A. A. Ewees (✉)
Computer Department, Damietta University, Damietta, Egypt
e-mail: a.ewees@hotmail.com

R. M. ElEraky · A. A. Ewees
Bisha University, Bisha, Kingdom of Saudi Arabia

R. M. ElEraky
Faculty of Specific Education, Damietta University, Damietta, Egypt

A. F. ElGamal
Computer Department, Faculty of Specific Education, Mansoura University, Mansoura, Egypt

© Springer Nature Switzerland AG 2020
M. Abd Elaziz et al. (eds.), *Recent Advances in NLP: The Case of Arabic Language*, Studies in Computational Intelligence 874,
https://doi.org/10.1007/978-3-030-34614-0_2

1 Introduction

The world wide web contains billions of documents and it is a rising at an exponential rapidity. Tools that supply timely access to, and digest of, numerous sources are vital in order to relieve the overloaded information people are meeting. These regards have sparked benefit in the expansion of automatic summarization systems [1]. The growing availability of online data has necessitated heavy research in the automatic text summarization area including the community of natural language processing (NLP) [2]. Document summarization is the method of generating an outline of a given document by reducing the size of the input text and saving its important data. There is a necessary to provide a summary with high quality and minimal time because currently the growth of information is growing tremendously on the web or on user's e-library, thus automatic document summarization (ATS) is the suitable tool to perform this task [3]. The purpose of text summarization is reducing the length and specifics of a document while keeping the most essential points and common meaning [4]. ATS is grouped to single document text summarization and multi-document summarization [5].

There are various related parameters, properties, and features that define several types or groups of text summarization. The basic parameters utilized in classifying text summarization are source or input documents (span) numbers, the document's languages numbers, summary specifics (summary length), objective audience, and summary structure [6, 7]. Unlike the achievements in English or different languages researches have not been completed appropriately on Arabic text summaries formatting. This is because of some problems and challenges that decreases the accuracy and achievements in Arabic NLP. These issues were inherited from the absence of automated Arabic NLP materials and the difficulty of the Arabic language. These obstacles can be summarized in the following points [8] (i) Arabic is extremely inflectional and derivational language, that creates morphological dissection like lemmatization and stemming a very complicated task, (ii) capitalization shortage in Arabic leading to a major challenge in the NER system operations, (iii) the lake of Diacritics named "Tashkeel" that is complementary in Arabic versions increases the inferring s' meaning complexity, (iv) and the absence of Arabic corpora in addition essential automated tools of Arabic NLP like lexicons, semantic role labelers, and the recognition of named entity makes the process more complicated. Besides, many challenges such as translation and summarization, one of the most serious semantic problems in translation and summarizing is the difference in the contextual distribution of words that appear to be synonymous in two languages, which may be synonymous with each other's, although they may differ in usage applications or language contexts [9]. It is also notable that grammatical confusion is a greater challenge when dealing with the Arabic language, and the previous studies of the automatic summary did not reach the accuracy of satisfactory to summarize the Arabic documents [10]. Compared to English document summarization, very few works were performed for Arabic document summarization [11].

Numerous methods are reported in previous studies for single document text summarization. These methods are grouped under a set of approaches which are semantic-based which is highly concerned with the concept of the terms as well as the connections/relations among terms, phrases, and clauses to construct the meant concepts of the text. Different semantic analysis mechanisms can be utilizes to summarize texts such as lexical chains and NLP methods [12–14], statistical based which is widely applied in summarizing contents. The meaning of relevance score which based on the extraction of a collection of features is the decisive operator that reflects the value of a clause regardless of its concept [15, 16], machine learning-based; in this approach, extractive content summarization operation is modeled as a dual classification issue. It based on a collection of statistical factors to train a dual classifier over a collection of training documents together with their human extractive outlines. Each clause in the document is symbolized as a vector of characteristics which are extracted from various levels; token, clauses, document and paragraph. The joint features among these levels based highly on word frequency, the position of the clause in the document or paragraph, the uniformity with the title, clause length [17–19], cluster-based that aims at classifying objects into groups drawing on the uniformity. In these approaches, the objects are the clauses, the classes are the clusters that the clauses belong to. The formation of the outline is executed by selecting a clause or more from every cluster based on the nearness to their cluster centroid [20, 21], graph-based; in this approach, the document is explained in a graph like the paradigm. In this paradigm, the nodes of the diagram represent the clause, while the links/edges among the linked nodes represent the uniformity relation among clauses. Therefore, a clause is considered significant if it is strongly linked to many other clauses. The implementation of this approach has a positive achievement in multi-document investigation communities [22–24], discourse-based summarization which is essential in defining the content conveyed by text. In this type, instead of handling the text like a continuity of terms and clauses, texts are organized like discourse-units are linked to each other to confirm discourse cohesion and coherence together. Building a successful discourse structures mostly rely on the availability of robust discourse parsers which based on four features including the pattern of discourse theory, the structure of data utilized for representing structure, the hierarchy and nature of the relations and lastly the language [25, 26], and finally an optimization-based approach; text summarization reflected by various researchers like a single or multi-objective optimization issues, where a collection of objectives considered producing a high-standard summary which contains coverage, diversity, coherence, and balance. Coverage means that summary should contain all important aspects appearing in the documents, while diversity aims to reduce the similar sentences in the output summary. Furthermore, coherence aims to generate a coherent text flow. Moreover, balance means that summary should have the same relative significant of different side of the original documents [27–29].

Recently, various document summarization mechanisms have been developed. Generally, those are containing the extractive or the abstractive mechanism. Extractive one generates the summary from sentences or phrases in the input content, in contrast, the abstractive one expresses the concept in the input content using various terms [3]. In [30] the authors proposed a model based on GA to set important features. The GA was utilized to learn each feature's weight by DUC 2002 dataset for training 100 documents. The ROUGE was used to check out the model by using the recall rate for a fitness function. Lastly, the results were compared with MS Word 2007 summarizer and Copernic summarizer together for 100 documents and 62 hidden documents. The authors of [31] proposed a novel method to generate a summary from Arabic text that based on SDRT. The method included two major parts. The first part creates the expatiate structure through extracting rhetorical connection among elementary discourse content and drawing SDRS graph that demonstrates the expatiate structure of the content. The second part builds the automatic outline depending on SDRS graph by reducing it throw removing rhetorical connections not supported in the outline chosen. The authors of [32] utilized the GA algorithm to increase readability out of clause cohesion via extracting clauses optimal combination. As well as, the informative features, they take into account various learner readability-regarding features like trigger terms percentage, average clause length, noun entity occurrences percentage and polysyllabic terms percentage for the summarization motivation. Their results found out that the extracted summary achieved the corpus-based approach and the baseline approach in terms of readability, measure, and cohesion.

Besides that, a new statistical summarization system was suggested in [11] for Arabic texts. That utilized a clustering technique and a modified discriminant analysis technique named MRMR to record words. The terms ranked based on their coverage power and discriminant. Also, it presented a novel clauses extraction technique, which selects clauses with top-ranked words and superior diversity. The suggested method applied lower language-dependent operations: sentence splitting, root extraction and tokenization. The outcomes presented that the suggested approach was competitive to other states of the art exists systems. The study [33] introduced a hybrid framework for the summarizing process to increase the outcomes of the QPM. The presented framework was based on RST and the network representation mechanism. The outcomes give good results in the summary. Froud et al. [34] investigated a validation for the impact of content summarization through the LSA model on Arabic Contents Clustering. They used five similarity scales which are Euclidean distance, Jaccard coefficient, cosine similarity, coefficient, Pearson correlation and averaged kullback-leibler divergence, for multiple times: with and without stopping. The experiment indicated that the suggested approach solves the noisy information and content length issues effectively, and consequently significantly enhance the clustering execution.

Jaradat and Al-Taani [35] proposed a mechanism to find a solution to the accuracy problem and the distraction of semantic relationships among the clauses issues of statistical-based methods using a scoring technique. Also, they solve the problem of Graph-based approaches in disregarding of clause structural features like its length, position. As well as investigating the effects of applying GA in Arabic single-document summarization extraction and solving the local optimum issue. The researchers of [36] used Genetic technique and the model of Map Reduce parallel programming for automatic text summarization of a big scale Arabic doubled documents. The suggested approach included scalability, rapidity and accuracy in generated summary. The suggested model outcomes showed that it eliminated clause redundancy and raised readability and cohesion operators among the clauses of summaries as well as the most important clauses identification. Also, in the study [37] a new generic abstract was presented for a single content in Arabic language summarizing. The framework first step segmented the input content topic wise. Then, the extractive summary from every textual segmented. Finally, apply rule-based clause reduction mechanism. The proposed summarizer was evaluated on a collection of 150 documents. The experimental outcomes showed noticeable enhancement in the execution, particularly the precision rate in shorter summaries.

2 Materials and Methods

Many researches handled Extraction-based approaches. On the contrary, the extractive-based has recently attract the attention of many researches. This paper relies on extractive summarization mechanism. The task of the extractive summarization is to find informative clauses, a subpart of a clause or phrase and include these extractive features into the final summary [38]. Figure 1 shows the prime steps of text summarization using extractive mechanism.

Stemming: is an aspect of the term which is utilized to create new terms via various linguistic mechanisms [40]. It should be noted that it is hard to determine the origin of any Arabic term since it needs a detailed syntactic, morphological, and semantic text analysis. Further, Arabic terms might not be derived from existing origin; they might have their own structures. In this chapter, the main task is to find out the origin of each term in text, since the root can be a base of various terms with informative related meaning. For example the root "play" is utilized for different terms relating to "playing" such as "player" and " ملعب". It is possible to find the Arabic origin automatically by removing the subparts of prefixes, suffixes, and infixes from the term [41]. Table 1. gives an example of the removal process. Thus, in this example the root of all the noted terms (العالم، معلمون، علوم) after removing subparts is the unique root of "elm" (علم).

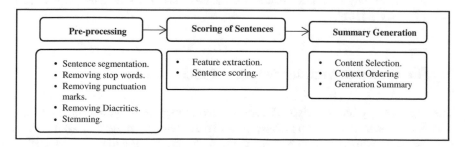

Fig. 1 The prime steps of text summarization using extractive mechanism [39]

Table 1 Various terms have different subparts and the identical root

Derivation (التفعيلة)	Suffixes	Infixes	Prefixes
العالم	-	١	ل+١
معلمون	و + ن	-	م
علوم	-	و	-

Frequency feature: Luhn presented a method relied on frequency. Frequency of a term has a pivotal role, to find out the importance of any term or clause in a given document. In our method the traditional method of "TF-IDF" measure was utilized which is defined as below, i.e. TF stands for the term frequency, IDF for reverse document frequency [42].

$$W_i = tf_i \times idf_i \tag{1}$$

$$idf_i = \log\left[\frac{D}{df_i}\right] \tag{2}$$

where, tf_i is the word frequency of ith word in the document, D represents the full number of documents, and idf_i is the document frequency of ith word in the full data set. In our implementation to calculate importance of word W_i for tf_i we considering the clause as a document and for IDF entire document as a data set [41].

Important terms: in the Arabic language, there are remarkable terms that increase the importance of the clause, such as: (this indicates that: ذلك يدلthe most important thing: اهم الأمور,...etc.). Such terms are saved in the database. Thus, the clause score increases if it has one or more of these terms according to the equation

Stop-word filtering: is a common mechanism used to counter the obvious fact that many of the terms included in a document do not contribute particularly to the description of the content of the documents [33]. Stop words are common terms that exist in the text but carry little meaning [43]. Remove all stop terms from sentences so that each clause has only the verbs and the nouns. It does not have a root, and it does not add any new information to the text. Examples of these terms are (هو ، هذا ، الذي ، هي) [41].

3 The Proposed Summarization Model

The suggested system consists of various phases beginning with entering of the Arabic document to be automatically summarized into the suggested system. The PHP language will be utilized to write the code for the suggested system as well as a set of other languages.

The proposed system composed of three main stages:

1. The pre-processing step.
2. The processing step.
3. The final summary step.

3.1 The Pre-processing Step

The pre-processing step aims to acquire a structured performance of the premier text, and this step includes some sub-steps as shown in Algorithm 1.

Algorithm 1: Preprocessing steps.

Input: all text
Operation:
- Split text using (. , ، ، ،)
- Remove iterative sentences.
- Remove parentheses and quotation marks ([ ] , ( ) , «» , { } , " ")
- Normalization alef by exchanging (ااا , اا , آ , إ , أ) to (ا) , Normalization yaa by exchanging (ي) to (ى) , and Normalization ta by exchanging (ة) to (ه).
- Removing any diacritics (ّ ، ِ ، ٍ ، ُ ، ً ، ٌ ، ْ ، َ ، ٌ)
- Removing any stop words from each sentence, stop words.
- Removing any punctuation from each sentence, punctuation marks like (: ، ؛ ،).
- Removing the subpart of suffixes (ات ، ون ، ين ، ان ، ها ، وا) from the word ending.
- Removing the subpart of prefixes (ال ، تال ، كال ، وال ، وكال ، وتال ، ولل ، لل) from the beginning of the word from each sentence by uniform expressions.
- Removing any spaces among words from each sentence.

Output: one array consists of all sentences.

3.2 The Processing Step

This stage contains the following points:

1. Calculate Sentence Feature: Each sentence is given a score, which serves as a good measure of the sentence by using a set of specific features. Each preset

features score takes a value ranging from (1, 0), the next set of features will be utilized:

- Frequency Feature: The weight of the sentence is computed on the basis of the recurrence of the term or synonyms and recurrence of relations by computing the mean rate of the frequency of the term in each clause as well as the synonyms. The weight of the term root is computed by Eq. (3) [41, 44] as below:

$$W_{i.j} = \log(N/n_i) * tf \tag{3}$$

where $W_{i.j}$ is the weight of the root in the sentence (j), N is the totality of the number of terms in paragraph, n_i the frequency of each term in the text, tf the duplication of the term and it is defined from $tf = \frac{n_i}{maxn_i}$ which means duplication of the term i divided by the highest repetition in the document.

Then determine the clause weight by Eq. (4) [41] as below:

$$S(i) = \sum (W_{i.j}) \tag{4}$$

where: S(i) is the weight of the clauses, the weights totality of the terms (i) in the clause (j).

Algorithm 2
Input: one array consists of clauses
Operation:
Set N to the number of clauses in the text
Set count_term to the number of terms in the text
For j = 1 to N
For i = 1 to count_term
Determine weight (Wi.j) for each term in the clause by Wi.j = log(N/ni) * tf
Determine weight S(j) for each clause by S(j) = sum (Wi.j)
Next term
Next clause
Output: S(j) (Repetition Feature)

- Important Words Feature: in Arabic there are some terms that are referred to significant information which can be contained in the summary like: (أهم الأمور، يدل ذلك) [33, 41]. Also in Arabic, the date (Hijri/Gregorian) is significant information that can be contained in the closing summary, a recent feature which has not been utilized before. and the score of the clause is computed after adding the significant terms through Eq. (5) [41] as below:

$$S(i) = \sum (W_{i,j}) + A \tag{5}$$

where: A is the significant terms in the text.

- The location feature: instituted a feature relies on sentence position. In [38] were almost manual but, later on this measure used widely in clause scoring, so leading clauses of an article are significant, and it takes a value among 0 and 1. The model utilized in this paper is given below, where N is the overall number of clauses. The utilized model is: $(Where\ 1 < i < N,\ and\ Score\ S(i) = [0, 1])$.

$$Score\ S(i) = 1 - \frac{i - 1}{N} \tag{6}$$

2. Sentence Scoring: in accordance with deducting the features of the clause, the weight of each clause is computed rely on the summation of Eqs. (5) and (6).

Algorithm 3
Input: one array consists of all clauses after stemming
Operation:
Compute all the features for all clause by counting 3 features (Repetition Feature, Position, Significant Terms)
Output: one array consists of Clauses Scoring

3.3 The Final Summary Step

The goal of this step is extracting the concluding summary over the following:

1. Rank calculation: The lowest weight of the clause that will be contained in the last summary is computed, by extracting the average rate of the clause weights by utilizing the next formula:

$$Rank = Round\left(\frac{\sum S(i)}{N}\right) \tag{7}$$

where: $\sum S(i)$ is a totality value of clauses' weights.

2. Last selection process: The higher-grade clauses (extracted from the former step) will be comprised in the last summary and the clauses with the minimum scores will be removed. The clauses will be comprised in the last summary as they occur in the primary text. In addition to the premier clause in case of (weight lower than the rank) because of its significance. This stage is done by Algorithm 4:

Algorithm 4: The last stage

Input: one array included clauses Scoring
Operation:
computing rank for each clause by Rank $= (\frac{\sum s(j)}{N})$
Check Frist clause
If (Frist clause score < Rank) then
      Summary = Frist clause
For i =2 to N
    If (clause scoring [i] > Rank) then
          Summary += '' + clause [i]
    End If
    Next
Output: summary (summary clause, as given in Source)

Figure 2 shows the steps of the proposed model whereas, Table 2 shows an example of a final summary.

4 Experiment and Results

In the suggested system, 33 articles were manually added one by one to summarize them and the summary ratios (per document) were computed by Eq. (8):

$$SR = \left(1 - \frac{\text{Number of terms in the summary (S)}}{\text{Number of terms in the document (D)}}\right) \times 100 \qquad (8)$$

4.1 Dataset

The research sample consisted of a set of articles collected from Wikipedia (https://ar.wikipedia.org/). Where a random sample of 33 articles in Arabic was selected in different fields including: astronomy, biology, chemistry, etc. The volume of articles

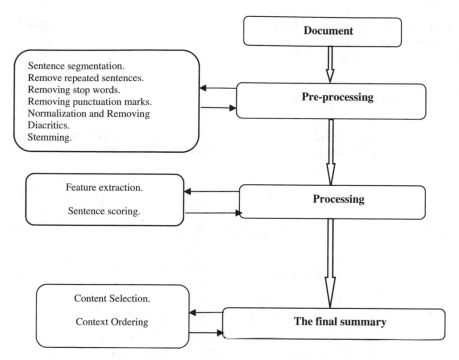

Fig. 2 Steps of the proposed model

Table 2 An example of a final summary

Original text
اللغة العربية هي أكثر اللغات تحدثاً ونطقاً ضمن مجموعة اللغات السامية، وإحدى أكثر اللغات انتشاراً في العالم، يتحدثها أكثر من ٤٢٢ مليون نسمة، ويتوزع متحدثوها في الوطن العربي، بالإضافة إلى العديد من المناطق الأخرى المجاورة، كالأحواز وتركيا وتشاد ومالي والسنغال وإرتيريا وإثيوبيا وجنوب السودان وإيران. اللغة العربية ذات أهمية قصوى لدى المسلمين، فهي لغة مقدسة (لغة القرآن)، ولا تتم الصلاة (وعبادات أخرى) في الإسلام إلا بإتقان بعض من كلماتها.
Generated summary
اللغة العربية هي أكثر اللغات تحدثاً ونطقاً ضمن مجموعة اللغات السامية، وإحدى أكثر اللغات انتشاراً في العالم، يتحدثها أكثر من ٤٢٢ مليون نسمة، ويتوزع متحدثوها في الوطن العربي، بالإضافة إلى العديد من المناطق الأخرى المجاورة،

ranged from long articles contains three or more paragraphs or a medium containing two paragraphs or a small paragraph containing one.

4.2 Evaluation Metrics

To evaluate the quality and efficiency of the suggested system, the following measures were used. These measures are used widely in several studies [45–47]:

$$\text{Recall} = \frac{\text{TP}}{\text{TP} + \text{FN}} \tag{9}$$

$$\text{Precision} = \frac{\text{TP}}{\text{TP} + \text{FP}} \tag{10}$$

$$\text{Fmeasure} = \frac{2 * \text{Recall*Precision}}{\text{Recall} + \text{Precision}} \tag{11}$$

$$\text{Accuracy} = \frac{\text{TP} + \text{TN}}{\text{TP} + \text{TN} + \text{FP} + \text{FN}} \tag{12}$$

where: TP is the totality of sentence pairs in the human expert summary and system summary. TN is the totality of pairs of sentences not found in the expert summary and system summary. FP is the totality of sentences in the system summary that are not in the expert summary. FN is the totality of sentences in the expert summary that are not in the system summary.

$$\text{Rouge} - \text{n} = \frac{\sum C \in \text{RSS} \sum \text{gram}_n \in C \, \text{Count}_{\text{match}} \, (\text{gram}_n)}{\sum C \in \text{RSS} \sum \text{gram}_n \in C \, \text{Count}(\text{gram}_n)} \tag{13}$$

where: N is the length of the N-gram, Count (N-gram) is the N-grams numbers that present in the reference summaries, and the ultimate number of N-grams co-occurring in the framework summary, the collection of reference summaries is Count match (N-gram) ROUGE measures mostly gives three main score Precision, Recall, and F-Score [38].

4.3 Evaluation of Summary Ratio

To judge the summary of the suggested system fairly, the articles were sent to two human experts to summarize the articles manually. Each expert summarized the data independently and recorded these summaries to be compared with the suggested system.

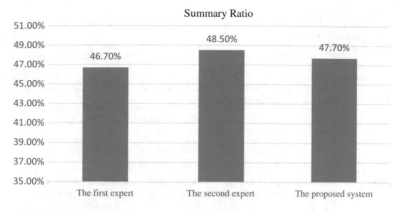

Fig. 3 Summary ratios of both experts versus the proposed model

The reduction ratio of the original documents by the proposed system has reached 47.7%, while the ratio of the first expert 46.7% and the second expert 48.5%. The differences in summary ratios are due to differences in the method of summarization between the proposed system and each of the two human experts. These results display the superiority of the suggested system to the human experts in the summary ratio. Figure 3 displays the summary ratio for each human expert compared to the proposed system.

4.4 Evaluation of the Proposed System Based on the Content

The proposed system is evaluated based on the content by human experts in the following four levels:

- The general form and content:
  To measure the output of the suggested system in the quality of the general form in terms of the order of paragraphs, sentences, the form of the beginning, the end of the texts, and measuring whether the content of the summarized text is appropriate in general, clear and understandable or that words without meaning.
- The coherence of the phrases:
  To measure the output of the suggested system in the consistency of the clauses, is there a correlation between them or not?
- Lack of elaboration or repetition:
  To measure the output quality of the suggested system in terms of lengthening in the summary, as well as measuring whether there are duplicate clauses.
- Completeness of the meaning:
  To measure the output quality of the suggested system in terms of whether the summary is complete or not

Table 3 The fifth Likert scale used to evaluate the four levels of the proposed system

Evaluation items	Refers to	Degree
Very poor	The sentence doesn't exist, the summary cannot be considered as representing the original text	5
Poor	Either the sentence is incomplete or illogical, or focuses on less important points in the original text	4
Fair	The sentence can be understood but needs effort, covering some important points in the original sentence	3
Good	The sentence can be easily understood and covers most of the important points that occur in the original text	2
Very Good	The sentence is read as if it were written by a human, that is, it covers the idea very well about what was discussed in the original text	1

Table 4 Estimated balance according to the fifth Likert scale

Response	Weighted average	General direction
Very poor	From 1 to 1.80	Very poor summarizing
Poor	From 1.81 to 2.60	Poor summarizing
Fair	From 2.61 to 3.40	Fair enough summarizing
Good	From 3.41 to 4.20	Good summarizing
Very Good	Higher than 4.20	Very good summarizing

These levels were evaluated using the fifth Likert scale, which contains five elements of assessment (very poor, poor, fair, good, and very good) as in Table 3. So that the number 1 as the relative weight expresses the lowest value and the number 5 as the relative weight expresses the highest value in the evaluation, as shown in the following:

An assessment scale was used to evaluate the questionnaire according to the fifth Likert scale as shown in Table 4:

The statistical results concluded that: The first statement "the general form and content": Statistical analysis shows that the general form and content obtained an arithmetic mean of 3.95, which is in the balance of the estimates of the fifth Likert scale: good, with a standard deviation 0.394. This indicates that the suggested system has been able to produce summaries of an appropriate general form, conceptually relevant content.

The second statement "The coherence of the phrases": As for the statistical analysis of the second axis, it is clear that the coherence of the phrases obtained an arithmetic mean of 3.97, which is in the balance of the estimates of the fifth Likert scale: good, with a standard deviation of 0.253. This indicates that the proposed system was able to produce summaries with coherent terms and a related meaning.

The third statement "Lack of elaboration or repetition": In the third axis, it obtained the highest arithmetic mean of 4.05, which is in the balance of the estimates of the fifth likert scale: good, with a standard deviation of 0.277. This indicates that the

Table 5 The arithmetical mean and the standard deviation of the assessment of human experts

Axis	The general form and content	The coherence of the phrases	Lack of elaboration or repetition	Completeness of the meaning
Arithmetic Mean	3.95	3.97	4.05	4.04
Standard Deviation	0.39	0.25	0.26	0.34

proposed system was able to produce good summaries without elaborating and does not contain repeated sentences.

The fourth statement "Completeness of the meaning": The fourth axis: obtained the arithmetic mean of the calculation of 4.04, which is in the balance of the estimates of the fifth Likert scale: good, and a standard deviation 0.335. This indicates that the suggested system was able to generate summaries of complete meaning, unbroken and content reader concept.

To summarize these results, the calculation of the arithmetical mean and the standard deviation of the assessment of human experts in the four axes of thirty-three articles were presented in Table 5.

As shown from the previous table:

(A) The human experts' evaluation degree of the suggested system (automatic summary) has generally been appropriate. The arithmetic average of the arbitrators' responses to the questionnaire was in the four axes as a whole 4.00.

(B) The "Lack of elaboration or repetition" axis came in the first order with the highest arithmetic mean 4.05, the "completeness of the meaning" axis in the second order with an arithmetic mean 4.04 and "The coherence of the phrases" axis come in the third with average 3.97. The "general form and content" axis came in fourth with an average of 3.94.

Figure 4 presents the statistical results of the expert responses in the four axes according to the pentagram of Carter.

For more details about this evaluation, Table 6 shows the evaluation's percentages of the human experts for each document in each level as well as the average of the responses are calculated for 33 documents and four levels.

From Table 6 the response's ratio for the human experts, in general, was in the four levels 80.15%. For example, the ratio of document 3 was 84.72%, which is the highest percentage in terms of general form and content was 81.40%, and the percentage of coherence of the phrases as well as completeness of the meaning was 85.40%, however the percentage of lack of elaboration or repetition reached 86.67% due to the uniqueness of the sentences, their clearly differences, and the use of various writing style which facilitated the process of determining the most significant sentence in the final summary.

The response's ratio for the human experts has decreased for document 31 by 71.95%, followed by the document 21 by 71.30% which is the lowest percentage, in

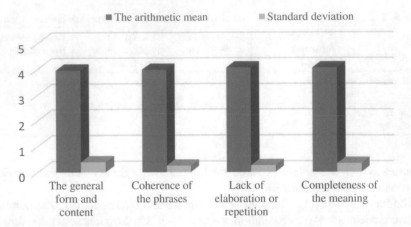

■ The arithmetic mean   ■ Standard deviation

Fig. 4 The statistical results of the experts' responses in the four axes according to fifth Likert scale

the terms of general form and content 73.40%, coherence of the phrases 70.60%, lack of elaboration or repetition 74.60%, and completeness of the meaning 66.60% due to non-writing of the documents scientifically, high similarities between the statements this caused some difficulties in determining the most significant sentence in the final summary. Besides, the system is based on deleting irrelevant sentences, not words.

4.5 Performance of the Proposed Method

To evaluate the performance of the suggested method five statistical measures have been used which are precision, recall, f-measure, accuracy rate, rouge. Table 7 represents the results of the five measures of the proposed method.

It is clear from Table 7 that the proposed system is superior to all previous systems and is ranked first on the Precession and F-measure scales of 72% and 69%, respectively; while in the second order according to the Recall scale of 68%; this result shows the efficiency of the proposed system in summarizing the articles compared to previous studies.

4.6 General Comparison Between the Proposed System and the Previous Studies

In order to judge the efficiency and the proposed system accuracy in the light of preceding studies, the results of the suggested system were compared with other automatic summary systems for Arabic texts. The following measurements were

Table 6 Evaluation's percentages of the human experts for each document and levels

Doc. No.	General form and content	Coherence of the phrases	Lack of elaboration or repetition	Completeness of the meaning	Average
1	81.40	78.67	86.67	86.67	83.35
2	77.40	85.40	86.67	85.40	83.72
3	81.40	85.40	86.67	85.40	84.72
4	78.60	81.40	84.00	84.00	82.00
5	76.00	81.40	84.00	81.00	81.00
6	80.00	84.00	84.00	85.00	83.00
7	85.00	81.00	78.60	85.00	82.40
8	74.60	73.40	82.60	77.40	77.00
9	77.40	81.00	82.60	81.00	80.50
10	80.00	85.40	80.00	79.00	81.10
11	77.40	80.00	80.00	80.00	79.35
12	84.00	83.00	83.00	79.00	82.15
13	81.40	77.00	83.00	79.00	80.20
14	79.00	80.00	85.40	83.00	81.85
15	79.00	85.40	85.40	86.60	84.10
16	81.40	81.00	84.00	81.00	81.85
17	84.00	81.40	81.40	82.60	82.00
18	78.60	78.60	82.60	78.60	79.60
19	74.60	78.60	78.60	82.60	78.60
20	72.00	78.60	77.40	79.00	77.00
21	73.40	70.60	74.60	66.60	71.30
22	80.00	78.60	76.00	79.00	78.00
23	79.00	77.40	80.00	76.00	78.10
24	79.00	74.40	77.40	82.60	78.35
25	80.00	84.00	81.40	82.60	82.00
26	80.00	77.33	76.00	80.00	78.00
27	76.00	74.60	83.00	83.00	79.00
28	83.00	81.40	80.00	84.00	82.10
29	77.00	80.00	83.00	74.60	78.65
30	83.00	77.00	80.00	81.40	80.35
31	68.00	73.00	77.00	69.00	71.95
32	77.00	81.40	79.00	76.00	78.00
33	86.60	80.00	80.00	84.00	82.65
Average	78.95	79.71	81.33	80.61	80.15

Table 7 Precision, Recall, F-measure, Accuracy, and ROUGE measures of the proposed method

Doc. Num	Prec.	Recall	F-measure	Acc.	Rouge
1	0.67	0.67	0.67	0.67	0.33
2	0.67	0.75	0.71	0.64	0.44
3	1	0.5	0.67	0.57	0.4
4	0.67	0.5	0.57	0.67	0.4
5	1	0.6	0.75	0.58	0.33
6	0.63	0.71	0.67	0.8	0.44
7	1	0.75	0.86	0.6	0.8
8	0.6	0.6	0.6	0.82	0.44
9	0.8	0.8	0.8	0.85	0.44
10	0.8	1	0.89	0.89	0.67
11	0.83	0.63	0.71	0.71	0.33
12	0.67	0.8	0.73	0.63	0.33
13	0.75	0.5	0.6	0.64	0.33
14	0.8	0.67	0.73	0.73	0.5
15	0.8	0.67	0.73	0.73	0.86
Avg	0.78	0.68	0.71	0.68	0.47

Table 8 Comparison between the arithmetic mean of the following measures: Recall, Precision, F-measure of the proposed system with some automated summary systems for Arabic texts

	Precision	Recall	F-measure
The proposed method	0.72	0.68	0.69
SDRT resume	0.56	0.85	0.65
R.I.A	0.69	0.42	0.60
ARST resume	0.63	0.45	0.50

used: Recall, Precision, F-measure, Accuracy, ROUGE1. It has been considered that ROUGE is an effective approach to measure document summarizes so greatly accept. ROUGE measures overlap terms between the system summary and standard summary (gold summary/human summary) [38].

Table 8 compares the arithmetic means of the following measures: Recall, Precision, F-measure of the proposed system with some automatic summary systems for Arabic texts [31, 48].

4.7 Statistical Analysis

To examine the imposition of the study, the ANOVA test was used using SPSS V22 program to detect whether there were statistically significant differences between the

Table 9 ANOVA analysis of the proposed system and the first and second expert

	Sum of squares	df	Mean square	F	Sig.
Between groups	61.057	2	30.529	0.345	0.709
Within groups	8485.582	96	88.391		
Total	8546.639	98			

automated summary (system summary) and the manual summary at 0.05 level in the sample studied.

It is clear from Table 9 that there are no statistically considerable differences among the summary of the system and summarizing both the first and second experts at the significance level of 0.05; where the value of F 0.345 at the significance level of 0.709 and is not statistically significant at the significance level of 0.05.

5 Conclusion

This paper suggested a model for the automatic summarization of Arabic texts. It was presented and discussed to summarize a single document. This model is based on the extractive method, where the summary contains of a set of important clauses from the premier text, and depends on the selection of clauses on the weight of each clause based on a collection of features, the processing is on the roots itself not the terms, and then the semantic similarity among the sentences is measured to select the most important clauses in the final summary after rearranging them in the identical order in the original text. The human evaluators 'evaluation of the proposed system was generally appropriate. The arithmetic mean of the arbitrators' responses to the questionnaire was in the four axes (general form and content, consistency of phrases, non- repetition, meaning completeness) according to the fifth Likert scale., which is a positive rate in favor of the proposed system, in addition to the superiority of the proposed system to the previous systems in the F-measure and ROUGE1 scale.

References

1. A. Nenkova, K. Mckeown, Automatic Summarization (USA, 2011), p. 1
2. S. Suneetha, automatic text summarization: the current state of the art. Int. J. Sci. Adv. Technol. **1**(9), (2011), ISSN: 2221-8386
3. R. Mol, Sabeeha: an automatic document summarization system using a fusion method. Int. Res. J. Eng. Technol. (IRJET), **3** (2016), ISSN: 2395-0056
4. Y. Rajput, P. Saxena, A combined approach for effective text mining using node clustering. Int. J. Adv. Res. Comput. Commun. Eng. **5**(4), 321–324 (2016), ISSN: 2319 5940
5. N. Bhatia, A. Jaiswal, Literature review on automatic text summarization: single and multiple summarizations. Int. J. Comput. Appl. (IJCA) **117**(6), 0975–8887 (2016)

6. D. Radev, S. Teufel, H. Saggion, W. Lam J. Blitzer A. Celebi, et al., Evaluation of text summarization in a cross-lingual information retrieval framework, (2011)
7. S. Lagrini, M. Redjimi, N. Azizi, Automatic arabic text summarization approaches. Int. J. Computer Appl. **164**(5) (2017)
8. A. Al-Saleh, M. Menail, Automatic Arabic text summarization: a survey. Artif. Intell. Rev. Arch **45**(2), 203–234 (2016)
9. M. Tafiqe, Y. Farag, M. Younis, Comparative and Contrastive Linguistics (Cairo University, 2014)
10. A. Basiony, *Computer for extracting knowledge and opinion mining* (Dar El Kotb El-elmia for publishing, Cairo-Egypt, 2011)
11. H. Oufaida, O. Noualib, P. Blache, Minimum redundancy and maximum relevance for single and multi-document Arabic text summarization. J. King Saud Univ.-Comput. Inf. Sci. 450–461 (2014)
12. K. Merchant, Y. Pande, NLP based latent semantic analysis for legal text summarization, in *2018 International Conference on Advances in Computing, Communications and Informatics (ICACCI)* (IEEE, 2018), pp. 1803–1807
13. A. Khan, N. Salim, H. Farman, M. Khan, B. Jan, A. Ahmad, A. Paul, Abstractive text summarization based on improved semantic graph approach. Int. J. Parallel Prog. **46**(5), 992–1016 (2018)
14. D.B. Patel,, S. Shah, H.R. Chhinkaniwala, Fuzzy logic based multi Document Summarization with improved sentence scoring and redundancy removal technique, Expert. Syst. Appl. (2019)
15. M.R. Chaud, A. Di Felippo, Exploring content selection strategies for multilingual multi-document summarization based on the universal network language (UNL). Revista de Estudos da Linguagem **26**(1), 45–71 (2018)
16. Cagliero, L., Garza, P., Baralis, E.: ELSA: a multilingual document summarization algorithm based on frequent itemsets and latent semantic analysis. ACM Trans. Inf. Syst. (TOIS), **37**(2) (2019)
17. S. Narayan, S.B. Cohen, M. Lapata, Ranking sentences for extractive summarization with reinforcement learning. arXiv preprint arXiv:1802.08636 (2018)
18. C. Kedzie, K. McKeown, H. Daume III, Content selection in deep learning models of summarization, arXiv preprint arXiv:1810.12343 (2018)
19. S. Song, H. Huang, T. Ruan, Abstractive text summarization using LSTM-CNN based deep learning. Multimed. Tools Appl. **78**(1), 857–875 (2019)
20. M.S. Bewoor, S.H. Patil, Empirical analysis of single and multi document summarization using clustering algorithms. Eng., Technol. Appl. Sci. Res. **8**(1), 2562–2567 (2018)
21. H. Van Lierde, T.W. Chow, Learning with fuzzy hypergraphs: a topical approach to query-oriented text summarization. Inf. Sci. **496**, 212–224 (2019)
22. P. Wu, Q. Zhou, Z. Lei, W. Qiu, X. Li: Template oriented text summarization via knowledge graph, in *2018 International Conference on Audio, Language and Image Processing (ICALIP)* (IEEE, 2018), pp. 79–83
23. Y. Wu, R. Chen, C. Li, S. Chen, W. Zou, Automatic summarization generation technology of network document based on knowledge graph, in *International Conference on Advanced Hybrid Information Processing*, (Springer, Cham, 2018), pp. 20–27
24. C. Mallick, A.K. Das, M. Dutta, A.K. Das, A. Sarkar, Graph-based text summarization using modified TextRank, in *Soft Computing in Data Analytics*, (Springer, Singapore, 2019), pp. 137–146
25. A. Cohan, N. Goharian, Scientific article summarization using citation-context and article's discourse structure. arXiv preprint arXiv:1704.06619 (2017)
26. X. Wang, Y. Yoshida, T. Hirao, K. Sudoh, M. Nagata, Summarization based on task-oriented discourse parsing. IEEE Trans. Audio Speech Lang. Process. **23**(8), 1358–1367 (2015)
27. R. Rautray, R.C. Balabantaray, Cat swarm optimization based evolutionary framework for multi document summarization. Phys. A **477**, 174–186 (2017)
28. J.M. Sanchez-Gomez, M.A. Vega-Rodríguez, C.J. Pérez, Extractive multi-document text summarization using a multi-objective artificial bee colony optimization approach. Knowl.-Based Syst. **159**, 1–8 (2018)

29. M.A. Mosa, A.S. Anwar, A. Hamouda, A survey of multiple types of text summarization based on swarm intelligence optimization techniques (2018)

30. L. Suanmali, N. Salim, M.S. Binwahlan, Genetic algorithm based sentence extraction for text summarization. Int. J. Innov. Comput. **1**(1), (2011)

31. Keskes, I., Lhioui, M., Benamara, F., Belguith, L.: Automatic summarization of Arabic texts biased on segmented discourse representation theory international computing conference in Arabic (ICCA, 26–28 December, Egypt 2012)

32. K. Nandhini, S.R. Balasundaram, Use of genetic algorithm for cohesive summary extraction to assist reading difficulties. Appl. Comput. Intell. Soft Comput. (2013)

33. F.G. El Sherief, Towards A Hybrid Framework for Automatic Arabic Summarizer, Unpublished Ph.D's thesis, Faculty of Computer and Information, Cairo University (2015)

34. H. Froud, A. Lachkar, S. Ouatik, Arabic text summarization based on latent semantic analysis to enhance arabic documents clustering. Colloq. Inf. Sci. Technol. (CIST) 22–24 October (2016)

35. Y.A. Jaradat, A.T. Al-Taani, Hybrid-based Arabic single-document text summarization approach using genatic algorithm, in *2016 7th International Conference on Information and Communication Systems (ICICS)*, (IEEE, 2016), pp. 85–91

36. R.S. Baraka, S.N. Al Breem, Automatic arabic text summarization for large scale multiple documents using genetic algorithm and mapreduce, in *2017 Palestinian International Conference on Information and Communication Technology (PICICT)*, (IEEE, 2017), pp. 40–45

37. A.M. Azmi, N.I. Altmami, An abstractive Arabic text summarizer with user controlled granularity. Inf. Process. Manage. **54**(6), 903–921 (2018)

38. Y.C. Shekhar, A. Sharan, Hybrid approach for single text document summarization using statistical and sentiment features. Int. J. Inf. Retr. Res. (IJIRR), 46–70 (2015)

39. Y.K. Menna, D. Gopalani, Feature priority based sentence filtering method for extractive automatic text Summarization (2015)

40. J. Singh, V. Gupta, A systematic review of text stemming techniques (2016)

41. A. Haboush, A. Momani, M. Al-Zoubi, M. Tarazi: Arabic text summarization model using clustering techniques. World Comput. Sci. Inf. Technol. J. WCSIT, **2**(3) 62–67 (2012)

42. M.M. Refaat, A.A. Ewees, M.M. Eisa, A.A. Sallam, Automated assessment of students' arabic free-text answers. Int. J. Intell. Comput. Inf. Sci. **12**(1), 213–222 (2012)

43. N. El-Fishawy, A. Hamouda, G. Attiya, M. Atef, Arabic summarization in Twitter social network. Ain Shams Eng. J. **5**(2), 411–420 (2014)

44. A.A. Ewees, M. Eisa, M.M. Refaat, Comparison of cosine similarity and k-NN for automated essays scoring. Cogn. Process. **3**(12) (2014)

45. R.A. Ibrahim, et al., Galaxy images classification using hybrid brain storm optimization with moth flame optimization. J. Astron. Telesc., Instrum., Syst. **4**(3), 038001 (2018)

46. E.H. Houssein, A.E. Ahmed, Mohamed Abd ElAziz. Improving twin support vector machine based on hybrid swarm optimizer for heartbeat classification. Pattern Recognit. Image Anal. **28**(2), 243–253 (2018)

47. M Abd Elaziz, A.A. Ewees, A.E. Hassanien, Multi-objective whale optimization algorithm for content-based image retrieval. Multimed. Tools Appl. **77**(19), 26135–26172 (2018)

48. M. Boudabous, M. Maaloul, I. Keskes, L. Belguith. Automatic summarization of arabic texts between digital learning theory and rhetorical structure theory. Commun. ACS, **4**(2) (2011)

A Proposed Natural Language Processing Preprocessing Procedures for Enhancing Arabic Text Summarization

Reda Elbarougy, Gamal Behery and Akram El Khatib

Abstract The techniques of pre-processing for text summarization process are utilized to get better text summarization performance. Previous researches have low performance concerning summarization of Arabic text graph-based procedure. This could be retrieved to: (1) Arabic is acomplicated language. (2) Dependence of the results in a graph-based procedure mainly on the weight between sentences. This paper discusses Arabic language techniques of pre-processing for text summarization process, with four major steps. These are tokenization, normalization, stopword removal and structural processing. Tokenization is to divide the text into its basic units. The text is separated into paragraphs, sentences, and words. Then, normalization processes take place with the following steps: removing diacritics, punctuations and duplicated white spaces, also then unifying character shapes. Stop-words are then deleted relying upon a pre-characterized list. In the final stage, morphological analyses and words stemming are done as a part of structural processing. In the basic procedure, words are labeled with their situation in sentences. This is important to get the number of nouns in the sentence sand the document to use as feature in Page Rank and minimum spanning tree weighing. Finally, Khoja stemmer return the best result in stemming while Alkhalil morphological analyzer returns the best results. The category of science and technology from EASC corpus is used as dataset to discuss the results of stop words removal, it returns better results than keeping stop words in the sentences.

Keywords Natural language processing · Features extraction · Morphological analyzer · Stop word · Text summarization

R. Elbarougy · G. Behery
Department of Computer Science, Faculty of Computer and Information Sciences,
Damietta University, New Damietta, Egypt
e-mail: elbarougy@du.edu.eg

G. Behery
e-mail: gbehery@du.edu.eg

A. El Khatib (✉)
Department of Mathematics, Faculty of Science, Damietta University, New Damietta, Egypt
e-mail: akram_elkhatib@hotmail.com

© Springer Nature Switzerland AG 2020
M. Abd Elaziz et al. (eds.), *Recent Advances in NLP: The Case of Arabic
Language*, Studies in Computational Intelligence 874,
https://doi.org/10.1007/978-3-030-34614-0_3

1 Introduction

This Paper explains the techniques of pre-processing used to enhance the results of text summarization. Syntax in Arabic language is very complex, so it's hard to deal with. Arabic language complexity comes from its complex morphology that allows the writer to switch between positions of words in the sentence, while retaining the meaning. Also, the structure of the word in Arabic is mind boggling, making standardization progressively troublesome. The names cannot be determined in the sentence because there are no uppercase lowercase letters. In addition, abandoning diacritics in the written text is another problem in Arabic which can influence the procedure of defining what the word position in this sentence is. The Arabic letter format varies based on the letter location in the word (beginning, middle or end of the word), which adds some complexity to the processing [1]. The number of nouns appearing in sentences means there is extra information in this sentence compared with other sentences. So, this approach focuses on using the power of using proper nouns in the feature extraction process and the weighing process [2], Therefore, to collect more information from the document you should take into your account the appearance of nouns [2]. Nouns are used in the process of calculating the weight of edges between nodes and the initial rank of the node in the graph based summarization system, such as page rank and minimum spanning tree algorithm.

Various stages take place on Arabic text to prepare it for the summary process. These processes are done to make the summarization process more accurate and easier. The operation of preprocessing is used to transform the Arabic text to be easily accessible in the approach of text summary. This stage contains methods, which remove diacritics, characters and even some words from the text. In general, preprocessing text may include the next processes: tokenization, normalization, deletion of stop-words, stemming, morphological analyses and Part-of-Speech (POS) tagging [3]. Then finally features needed in the summarization process are extracted. For example, nouns are extracted depending on the results of part of speech tagging. Fig. 1 shows these steps:

2 Literature Review

Since the second half of the twentieth century, many researches have been done on text summary. In this regard, different summarization techniques were classified into various grouping on specific features of each approach.

Hovy and Lin [4], define a text summary as a text based on a document or more. This summary contains the most important information of the main texts and its content is to the extent of half of the main texts. Mani and Maybury [5], defines the text overview as "a method of discovering the primary source of data, discovering important material, and displaying it as a concise document".

Earlier research in text summarization was done by using statistical features of text only to calculate the importance of sentences. Douzidia [6], is one of the scientists who suggested a statistical system of text summarization using distinct requirements to calculate the weight of sentences. The pre-processing methods used here involve normalization, tokenization and withdrawal of stop-words based on a particular list and lemmatization. El-Haj [7], is another researcher who proposed a multi-document extractive text summarization model containing four stages to extract Arabic text summary. The proposed steps are as follows: Get data, tokenization, stemming and delete of stop-words, and feature extraction, scoring, ranking and generating the final summary.

Another approach was released which is by using linguistically processing methods. As suggested Hadni et al. [8], proposed a combination of linguistics and a statistical approach used as a hybrid technique for extraction of Arabic multiword term. The researcher uses text segmentation and speech part tagging as techniques of pre-processing. Sawalha and Atwell [9], have have suggested a technique that utilizes distinct kinds of morphological assessment in Arabic. They found that Khoja stemmer was more accurate than other analyzers in the words having roots containing three characters. 80–85% of Arabic words are formed of a three-letter root, while the rest of the words are formed of four, five and six-letter roots [10]. So here, this research use Khoja stemmer because as it achieves the highest accuracy rate for words of three-letter roots and also performed well in four-letter roots.

Haboush et al. [11], proposed a model for summarizing Arabic text using techniques for clustering. The techniques of pre-processing used in this research start from tokenization, dividing the text into paragraphs, into sentences, then into words. After that, stop words are removed and stemming finally applied.

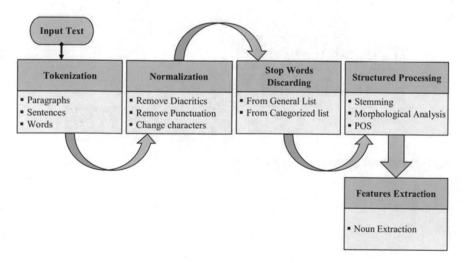

Fig. 1 Text preprocessing stages

Finally, according to the previous researches the text summarization pre-processing start with the process getting data, tokenization, normalization and stop-words removal. Some researches use morphological and structural analysis.

The process of Arabic text summarization can go through many steps, starting with techniques of text pre-processing such as normalization, stemming and morphological analyzing, and then features extraction, followed by applying scoring methods and finally extracting the summary.

3 Preprocessing for Arabic Text Summarization

In this section, we discuss our method used for Arabic text pre-processing, which have been applied successfully to different NLP Processing tasks. It includes the following processes: tokenization, normalization, deletion of stop-words, stemming, morphological analysis and tagging of the POS.

3.1 Tokenization:

In the stage of text pre-processing, tokenization process or segmentation is the first process to do. Tokenization, is the process of splitting text into its basic units. This unit may be paragraphs, sentences, words, characters, numbers or any other appropriate unit [12]. The main basic units in tokenization are paragraph, sentence, and words [13]. Fig. 2 shows the process of tokenization.

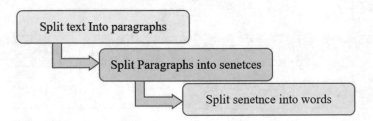

Fig. 2 Tokenization process

Algorithm 1: Tokenization Algorithm.

```
Input: Document
Output: List of Word Tokens
Split Document into Paragraphs
Foreach Paragraph in Document do
        Split Paragraph into Sentences
        Foreach Sentence in Document do
                Split Sentence into Tokens
        End
End
```

Algorithm 1 shows the process of tokenization that starts from splitting the document to paragraphs, then splitting each paragraph to sentences; and finally splitting each sentence to its basic tokens.

Depending on the morphological complexity of the Arabic language, Arabic tokenization suffers from some restrictions: (1) There is no capital character defining the beginning of the word. (2) Complex words in Arabic connect two words together like "Abdullah". (3) The absence of punctuation makes it difficult to determine the end of sentences. Tokenizer has two main issues: (1) Splitting text into its tokens. (2) Definition of unit boundaries. Tokenization steps start from splitting text into paragraphs, sentences and words as presented in Fig. 2:

1. *Split Text into Paragraphs*: In this step the text is split into paragraphs. Table 1 shows an example of a text containing 3 paragraphs before and after splitting.
2. *Split Text or Paragraph into Sentences*: The problem in the processing of natural languages is where sentences start or end. Table 2 shows an example of a three-sentence and the splitting process.
3. *Split Sentences into basic units*: The most basic unit is the word. The boundaries to select the word are different when they are at the beginning, center or end of the sentence as shown in Fig. 3. Table 3 shows the process of splitting a sentence to its basic tokens.

3.2 Normalization:

Normalization is the process of making a text more consistent by replacing some characters, removing duplicate white spaces, removing diacritics and so on. In general, the process of normalization is divided into two categories: removing and replacing, as in Fig. 4. According to Fig. 4 normalizations is done as follows:

1. *Removing*: In this step characters, punctuation and diacritics are removed to make the processing of text summarization easy. Punctuation is used for organizing the Arabic text and making it more readable to give the actual meaning to the reader. Table 4 shows the list of punctuations to be removed. In addition, Table

5 shows an example of a sentence before and after the process of removing punctuation. In addition, another thing to remove is diacritics. Table 6 and Table 7 shows an example of removing diacritics. Moreover, duplicate white spaces make the process of tokenization return empty words, which is wrong. To avoid this, to minimize problems all white spaces are replaced with one white space. The Defining Article " الـ " in Arabic is used to convert the Arabic words from the undefined form to the defined form. It corresponds to article "the" in English language. Table 8 shows some forms of " الـ " in Arabic language. Table 9 shows an example of removing the defining article. The benefit from removing the defining article is to provide Arabic words with same syntax to be considered as one word in the process of calculating term frequency feature of the sentence. Non-letter characters may not be helpful in the process of summarization. Therefore, numbers and digits are removed from the text like history, times and maybe

Table 1 Document to paragraph segmentation

Text	التمويل (Finance): يعنى التمويل بتحديد احتياجات الأفراد والمنظمات والشركات من الموارد النقدية وتحديد سبل جمعها واستخدامها مع الأخذ في الحسبان المخاطر المرتبطة بمشاريعهم. عليه فإن مصطلح تمويل يجمع بين التالي: دراسة النقود وغيره من الأصول، إدارة هذه الأصول ورقابتها، تحديد مخاطر المشاريع وإدارتها، علم إدارة المال. في صيغة الفعل فإن كلمة تمويل تعني توفير الاعتمادات المالية للأعمال أو للمشتريات الفردية الضخمة (مثل السيارات والمساكن). ويعمل المصرف على تجميع أنشطة الدائنين والمقترضين. فيقبل المصرف ودائع من الدائنين يدفع عنها فائدة معلومة، ومن ثم يقوم بإعارة هذه الودائع للمقترضين. هكذا فإن المصرف يقوم بتنسيق أنشطة الدائنين والمقترضين مع بعضهم البعض مهما اختلفت احجامهم، ويمكن القول إن المصرف يمنع تدفع النقود في الهواء.	
Paragraph Tokenization	Paragraph 1	التمويل (Finance): يعنى التمويل بتحديد احتياجات الأفراد والمنظمات والشركات من الموارد النقدية وتحديد سبل جمعها واستخدامها مع الأخذ في الحسبان المخاطر المرتبطة بمشاريعهم. عليه فإن مصطلح تمويل يجمع بين التالي: دراسة النقود وغيره من الأصول، إدارة هذه الأصول ورقابتها، تحديد مخاطر المشاريع وإدارتها، علم إدارة المال.
	Paragraph 2	في صيغة الفعل فإن كلمة تمويل تعني توفير الاعتمادات المالية للأعمال أو للمشتريات الفردية الضخمة (مثل السيارات والمساكن).
	Paragraph 3	ويعمل المصرف على تجميع أنشطة الدائنين والمقترضين. فيقبل المصرف ودائع من الدائنين يدفع عنها فائدة معلومة، ومن ثم يقوم بإعارة هذه الودائع للمقترضين. هكذا فإن المصرف يقوم بتنسيق أنشطة الدائنين والمقترضين مع بعضهم البعض مهما اختلفت احجامهم، ويمكن القول إن المصرف يمنع تدفع النقود في الهواء.

Table 2 Paragraph tokenization example

Paragraph		ويعمل المصرف على تجميع أنشطة الدائنين والمقترضين. فيقبل المصرف ودائع من الدائنين يدفع عنها فائدة معلومة، ومن ثم يقوم بإعارة هذه الودائع للمقترضين. هكذا فإن المصرف يقوم بتنسيق أنشطة الدائنين والمقترضين مع بعضهم البعض مهما اختلفت احجامهم، ويمكن القول إن المصرف يمنع تدفع النقود في الهواء.
Sentences	Sentence 1	ويعمل المصرف على تجميع أنشطة الدائنين والمقترضين.
	Sentence 2	فيقبل المصرف ودائع من الدائنين يدفع عنها فائدة معلومة، ومن ثم يقوم بإعارة هذه الودائع للمقترضين.
	Sentence 3	هكذا فإن المصرف يقوم بتنسيق أنشطة الدائنين والمقترضين مع بعضهم البعض مهما اختلفت احجامهم، ويمكن القول إن المصرف يمنع تدفع النقود في الهواء.

Table 3 Sentence Tokenization Example

Sentence							فيقبل المصرف ودائع من الدائنين يدفع عنها فائدة معلومة، ومن ثم يقوم بإعارة هذه الودائع للمقترضين.
Tokens	فائدة	عنها	يدفع	الدائنين	من	ودائع	المصرف	فيقبل
	للمقترضين	الودائع	هذه	بإعارة	يقوم	ثم	ومن	معلومة

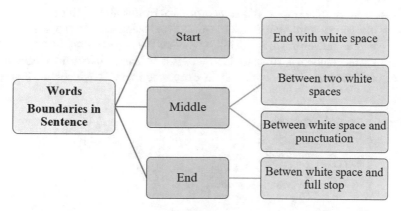

Fig. 3 Words boundaries

page numbers and so on from the text. Table 10 shows an example of removing non-letter characters from the text. Non-Arabic words or scientific glossaries are usually written in English. These words are not important in the process of text summarization. Therefore, in this step, these words are removed. Strange words are the words with non-Arabic origins, as defined in the process of morphological analysis. Also it can be removed according to predefined list. Table 11 shows an example of non-Arabic words being removed from text as in Table 12 or algorithm. Table 13 shows an example of removing a strange word from the text.

2. *Letters Normalization*: Arabic letters have different shapes depending on their position in words or their diacritics. So some Arabic letters are unified in normalization to enhance the process of term weighing, extraction features, and semantic features. There are three letters in Arabic that need to normalize as follows: Check and Convert ALEF Letter: ALEF is the first letter in Arabic alphabet. This letter is written in 4 different forms depending on its position in the sentence as denoted in Table 14. The system converts all these different forms into one " ا " form to help stemmer process. Table 15 shows an example of normalization of the ALEF character. And check and convert "ى " and "ة" characters: both letters are normalized in this step as in Table 16.

3. *Stop Words Discarding*: Stop-words frequently occur, insignificant words appear in an article or web page, which is non-informative. In addition, they shorten the document's length, thus affecting the weighing process [14]. Moreover, they do not help identifying document topics and reducing features [15]. Table 17 shows examples of Stop-words.

There is no definite stop-word list incorporated by that all NLP tools. Not all NLP tools are using a word-stop list. Some tools avoid using these to support searching for phrases. However, when deciding whether the current word is stop-word or not such as numbers in some topics, there is a difference in consideration, it will be considered as a stop-word, but in economic topics it will be an important

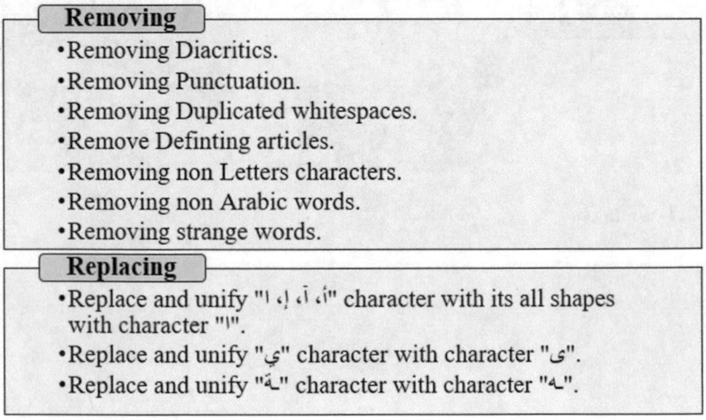

Removing
- Removing Diacritics.
- Removing Punctuation.
- Removing Duplicated whitespaces.
- Remove Definting articles.
- Removing non Letters characters.
- Removing non Arabic words.
- Removing strange words.

Replacing
- Replace and unify "ا ، إ ، آ ، أ" character with its all shapes with character "ا".
- Replace and unify "ى" character with character "ى".
- Replace and unify "ة" character with character "ه".

Fig. 4 Arabic Text Normalization

Table 4 Punctuations list

'	.	>	<	(	)
$	%	^	&	!	/
_	-	=	+	*	\
؛	؟	،	'	÷	×
~	`	:	"	{	}
|	،	@	#	[	]

Table 5 Punctuation removal example

Before	فيقبل المصرف ودائع من الدائنين يدفع عنها فائدة معلومةً، ومن ثم يقوم بإعارة هذه الودائع للمقترضين.
After	فيقبل المصرف ودائع من الدائنين يدفع عنها فائدة معلومة ومن ثم يقوم بإعارة هذه الودائع للمقترضين

word. Another example is dates that are important in historical topics. Algorithm 2 displays the process of removing stop-words.

Algorithm 2: Stop Words Removal Algorithm.

```
Input: Document
Output: Sentences without Stop Words
For Each Sentence in Document
    For Each Word in Sentence
        If (Word is Stop Word )
            remove();
        End If
    End For
End For
```

Stop-words can be categorized into many categories [14]: adverbs, units of measurement, coin names, conditional pronouns, interrogative pronouns, prepositions, pronouns, reference names/determinants, relative pronouns, transformers (verbs, letters), and verbal pronouns etc. In Arabic, stop words can be classified into two categories to extend them to prefixes or suffixes: Words can take suffixes or prefixes. Table 18 illustrates an example of stop words that may take prefixes or suffixes. And Words can't take suffixes or prefixes like this (ثُم ، أو). All forms of stop-words with suffixes and prefixes are taken into account and added to the list of stop-words. Table 19 shows an example of the deletion of stop-words.

Table 6 Diacritics list

ó	ó	ọ	ó
ó	ó	ọ	ó

Table 7 Diacritics removing example

Before	أَدْخُلِ الْبُسْتَانَ وَمُتِّعْ نَفْسَكَ بِمَنْظَرِهِ، وَلَا تَعْبَثْ بِأَزْهَارِهِ وَثِمَارِهِ.
After	ادخل البستان ومتع نفسك بمنظره، ولا تعبث بأزهاره وثماره.

Table 8 Arabic defining article forms

Arabic Defining Article					
ال	وال	كال	فال	بال	ل

Table 9 Removing defining article example

Before	ويعمل المصرف على تجميع أنشطة الدائنين والمقترضين.
After	ويعمل مصرف على تجميع أنشطة دائنين مقترضين.

Table 10 Removing non-letter character example

Before	فإذا كنت تملك سهم واحد من الشركة س والتي يبلغ مجموع اسهمها 100 سهم فإنك بهذا تمتلك 1% من الشركة.
After	فإذا كنت تملك سهم واحد من الشركة س والتي يبلغ مجموع اسهمها سهم فإنك بهذا تمتلك من الشركة.

Table 11 Removing non-arabic words example

Before	التمويل (Finance): يعنى التمويل بتحديد احتياجات الأفراد والمنظمات والشركات من الموارد النقدية وتحديد سبل جمعها واستخدامها مع الأخذ في الحسبان المخاطر المرتبطة بمشاريعهم.
After	التمويل : يعنى التمويل بتحديد احتياجات الأفراد والمنظمات والشركات من الموارد النقدية وتحديد سبل جمعها واستخدامها مع الأخذ في الحسبان المخاطر المرتبطة بمشاريعهم.

Table 12 Strange words example

معاريف	هأرتس	فرانس	الغارديان	الناتو
سوروكا	يونايتد	سيتي	برس	لاتيه

Table 13 Strange words removal example

Before	كما أكد مصدر مطلع من حلف **الناتو** لصحيفة **الغارديان** عن قلقه العميق من التوترات التي تجري في منطقة الشرق الأوسط.
After	كما أكد مصدر مطلع من حلف لصحيفة عن قلقه العميق من التوترات التي تجري في منطقة الشرق الأوسط.

Table 14 LEF shapes

ALEF Different Shapes		ALEF Normalized Shape
أ	ا	ا
إ	آ	

Table 15 Example of ALEF letter normalization

Before	الناي آلة نفخية تعد بحق أقدم آلة موسيقية في التاريخ إذا استثنينا الآلات الإيقاعية
After	الناي الة نفخية تعد بحق اقدم الة موسيقية في التاريخ اذا استثنينا الالات الايقاعية

Table 16 Letter normalization

Letter Shapes		Normalized Shapes
ي	ى	ى
ة	ه	ه

Table 17 Stop words list

من	إلى	على	كلا	كان	إن
كيف	عام	ميلادي	في	أمام	إذا

Table 18 Stop words with suffixes and prefixes

Stop Word	With Suffix	With Prefix	Conjugation / Article
كان	بكونهم	وكان	يكون
أول	أولهم	بأول	الأول

Table 19 Stop words removal example

Sentence	هكذا فإن المصرف يقوم بتنسيق أنشطة الدائنين والمقترضين مع بعضهم البعض مهما اختلفت احجامهم، ويمكن القول إن المصرف يمنع تدفع النقود في الهواء.
Stop Words Re-moval	المصرف يقوم بتنسيق أنشطة الدائنين المقترضين اختلفت احجامهم، القول المصرف يمنع تدفع النقود الهواء.

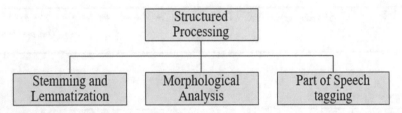

Fig. 5 Structured processing categories

3.3 Structured Preprocessing:

Some structures and NLP are used in this process to enhance the text summary results. As shown in Fig. 5, the structured process can be divided into three major aspects; stemming, morphological analysis and part of speech tagging.

3.4 Stemming and Lemmatization:

In the summary process, a basic unit is needed to be used in the summary, as the basic unit is the smallest unit used in the summarization process. As shown in Fig. 6, this basic unit may have different type. We've got the following basic units:

1. *No-stem*: The words used in this basic unit as found without any stemming or lemmatization.
2. *N-gram*: In n-gram, the text is divided into chunks of size N and this gram is used as the basic unit in the summary process [16, 17]. Table 20 shows an n-gram example for "الشركات". Algorithm 3 shows the n-gram generation algorithm.
3. *Stem*: Stemming is the process of getting the word stem in Arabic language. Stemming is done by removing any affixes attached to the Arabic words. For example, in statistical text analysis, stemming helps map grammatical variations of a word to instances of the same term [18–20]. There is a major approach followed for Arabic stemming: Rooted Stemmer: In this stemmer, we use Khoja stemmer [19], that stems the word to its root form that has three or four letters. Rooted stemmer can be done by two approaches; statistical stemmer or manually

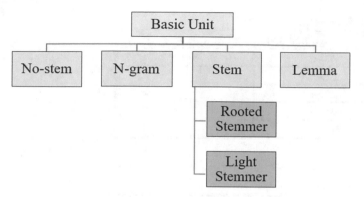

Fig. 6 Basic units

Table 20 N-gram example

N-gram for "الشركات" where n = 3				
كات	ركا	شرك	لشر	الش

built dictionary stemmers [21]. The manually built dictionaries that have the roots of all Arabic words are called lemmatization [22]. Tables 21, 22, and 23, shows an example of rooted stemmer.

4. *Lemma*: Lemmatization is the process of getting the root of Arabic words through some morphological analysis. In order to perform lemmatization, we need to have some predefined dictionaries to map between the words and their roots: So, which is better stemming or lemmatization, we can speak here according to two keys: the first is the time needed to develop, and the other key is the accuracy of the results as summarized in Table 24, It's easier to develop a stemmer than to build a lemmatizer. Because in lemmatizer, a deep linguistic knowledge is needed to create dictionaries that allows the algorithm to search for the proper form of the word. However, the result of the lemmatizer will be more accurate, depending on the accuracy of the results provided on the information retrieval process. In this research statistical stemmer is used because it is faster than lemmatization and has acceptable accuracy.

```
Algorithm 3: N-gram Generating Algorithm.
Input: Text, N=n
Output: N-gram List
Index=0
N_gram_list
While index+2 < Length(Text) then
        N_gram_list.add(text.subString(index,N))
        Index= index+1
End While
```

Table 21 Rooted stemmer example

Rooted stemmer Example	
Word	Root
الدرس	
يدرس	درس
مدرس	
مدرسة	

Table 22 Word affixes example

Antefix	Prefix	Core	Suffix	Postfix
ل	ي	ناقش	و	هم

Table 23 Light10 stemmer example

Words	Light10 Stem
تطورا	تطور
أسماء	اسماء
الماضي	ماض
الشارقة	شارق

Table 24 Comparison between stemming and lemmatization

	Stemming	Lemmatization
Developing Speed	Fast	Slow
Accuracy	Acceptable	High

3.5 Morphological Analysis

Arabic is one of the most complex morphological languages. There are no capital and small letters in Arabic as an example, so we cannot distinguish between nouns and non-noun words. Another unique problem in Arabic sentence is that the sentence can either begin with a name or a verb. Arabic language morphology is based on the basic pattern of word formation. Therefore, most native Arabic words are derived from basic entities called roots or stems. Roots are radicals relying on a list of predefined models called morphological models. Every Arabic term is created either by using its root or by incorporating suffixes or prefixes to its core. Each word in the sentence has its own POS that matches its position in the sentence and the connection between the sentences and the sentence root. There is many Arabic morphological analysis such as Buckwalter Arabic Morphological Analyzer (BAMA) [23], and Alkhalil Morpho [24].

3.6 Part of Speech Tagging (POS)

POS tagging is the process of giving a syntactic role to each word in context and is therefore considered a crucial step that greatly affects other subsequent NLP tasks. Traditionally, Arabic grammar does the analysis of all Arabic words to three main POS. These POS are further sub-categorized into more detailed parts of speech that collectively cover the entire Arabic language [3]. The three main POS's: Noun, which is a name or a word describing a person, a thing or an idea. Traditionally, the noun classes in Arabic are divided into: nouns derived from verbs, nouns derived from other nouns, nouns derived from particles and primitives. Verb, the verb in Arabic is similar to that in English, although the tenses and aspects are different. There are some suffixes added to verbs depending on their tenses or the number of subjects or their gender. As for particles, these include prepositions, adverbs, conjunctions, interrogative particles, and interjections. There are many systems and libraries in Arabic that are used in the tagging process [25, 26].

3.7 Features Extraction (Names)

Arabic nouns are words describing a person, a thing, a place, or an idea. When nouns appear in sentences, the sentence weight is increased with presence of extra information in this sentence compared to other sentences. Therefore, this approach is focused on the power of using proper nouns in the process of feature extraction and weighing process [2]. To extract the nouns from the text. Morphological analysis is applied, then POS tagging is also used. Table 25 provides an example of Alkhalil morphological analysis noun extraction [24].

4 Experimentation and Results

4.1 Dataset (Corpus)

To evaluate the proposed approach, the Essex Arabic Summaries Corpus (EASC) is used [27]. EASC includes 10 subjects: art, music, environment, politics, sports, health, finance, science and technology, tourism, religion, and education

Table 25 Nouns extraction

Paragraph No	Sentence No	Sentence	# of Nouns
01	01	التمويل (Finance): يعنى التمويل بتحديد احتياجات الأفراد والمنظمات والشركات من الموارد النقدية وتحديد سبل جمعها واستخدامها مع الأخذ في الحسبان المخاطر المرتبطة بمشاريعهم.	6
	02	عليه فإن مصطلح تمويل يجمع بين التالي: دراسة النقود وغيره من الأصول، إدارة هذه الأصول ورقابتها، تحديد مخاطر المشاريع وإدارتها، علم إدارة المال.	7
02	01	في صيغة الفعل فإن كلمة تمويل تعني توفير الاعتمادات المالية للأعمال أو للمشتريات الفردية الضخمة (مثل السيارات والمساكن).	7
03	01	ويعمل المصرف على تجميع أنشطة الدائنين والمقترضين.	3
	02	فيقبل المصرف ودائع من الدائنين يدفع عنها فائدة معلومة، ومن ثم يقوم بإعارة هذه الودائع للمقترضين.	6
	03	هكذا فإن المصرف يقوم بتنسيق أنشطة الدائنين والمقترضين مع بعضهم البعض مهما اختلفت احجامهم، ويمكن القول إن المصرف يمنع تدفع النقود في الهواء.	6

4.2 Evaluation Metrics:

The evaluation calculates with respect to Precision, Recall and F-measure. The value of Precision recall and F-measure will be calculated as in Formula (1), Formula (2) and Formula (3) respectively.

1. *Precision*: To metric the correct text size that is returned by the system.

$$Precision = \frac{Extracted\ Summary \cap Provided\ Summary}{Extracted\ Summary} \qquad (1)$$

2. *Recall*: The metric coverage system reflects the ratio of the extracted relevant sentences.

$$Recall = \frac{Extracted\ Summary \cap Provided\ Summary}{Provided\ Summary} \qquad (2)$$

3. *F-measure*: Works a balance relation among recall metric and precision metric.

$$F\text{-}measure = \frac{2 * Precision * Recall}{Precision + Recall} \qquad (3)$$

Table 26 Summary metrics using stop words

Essex arabic summaries corpus (EASC) category	Precision	Recall	F-measure
Science and technology category - with stopword	65.88	66.68	63.74
Science and technology category - without stopword	73.86	69.67	69.36

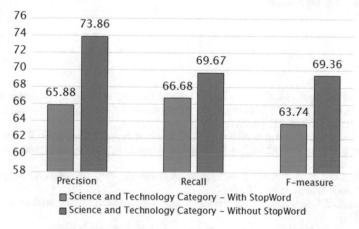

Fig. 7 Summary metrics using stop words

4.3 Experiment Setup:

In this section science and technology category is used from EASC corpus to evaluate the system. The results done using AL Khalil morphological analyzer [24]. The effect of stop words is discussed here. The experiment is conducted with and without using stop words removal. Table 26 and Fig. 7 show the results of using stop words removal which obviously show that stop words removal enhance the performance of the summary.

As explained previously stop words make more calculation without any important in the sentences so removing it improves the results of summary metrics.

5 Conclusion

This Paper discussed the stage of pre-processing, which may also be recognized as a part of NLP. The techniques of pre-processing are very important in Arabic language due to the complexity of its structure. The techniques of pre-processing were orderly presented and exemplified. These pre-processing techniques are used to improve Arabic text summary performance. Pre-processing techniques include tokenization, standardization, stop-words removal and structural analysis. Nouns are extracted from the text as a result of structural analysis. The number of nouns in the sentence is used as a feature to improve the quality of text summary. It is not necessary to apply all these steps. What you need is only to apply the steps that will be helpful to your algorithm. The category of science and technology from EASC corpus is used as dataset to discuss the results of stop words removal, it returns better results than keeping stop words in the sentences with 73.86% for precision, 69.67% for recall and 69.36% for f-measure.

References

1. N. Alami, M. Meknassi, S.A. Ouatik, N. Ennahnahi, Impact of stemming on Arabic text summarization, in *2016 4th IEEE International Colloquium on Information Science and Technology (CiSt) IEEE*, pp. 338–343 (2016)
2. R. Al-Shalabi, G. Kanaan, B. Al-Sarayreh, K. Khanfar, A. Al-Ghonmein, H. Talhouni, S. Al-Azazmeh, *Proper Noun Extracting Algorithm for Arabic Language* (In International conference on IT, Thailand, 2009)
3. J.A. Haywood, H.M. Nahmad, G.W. Thatcher, *A New Arabic Grammar of the Written Language* (Lund Humphries, London, 1965)
4. E. Hovy, C.Y. Lin, Automated text summarization in SUMMARIST. Adv. Autom. Text Summ. **14** (1999)
5. I. Mani, M.T. Maybury, Automatic summarization John Benjamin's publishing Co (2001)
6. F.S. Douzidia, *Automatic Summarization of Arabic Text*. (University of Montreal, Memory presented at the Faculty of Graduate Studies, 2004)
7. M. El-Haj, Multi-document arabic text summarisation (Doctoral dissertation, University of Essex, 2012)
8. Hadni, M., Lachkar, A., Ouatik, S.A.: Multi-Word Term Extraction based on New Hybrid Approach for Arabic Language (2014)
9. M. Sawalha, E. Atwell, Comparative evaluation of Arabic language morphological analysers and stemmers. Coling 2008: Companion volume: Posters, 107–110 (2008)
10. S. Eldin, *Development of a Computer-Based Arabic Lexicon* (In the Int. Symposium on Computers & Arabic Language, ISCAL, Riyadh, KSA, 2007)
11. Haboush, A., Al-Zoubi, M., Momani, A., Tarazi, M.: Arabic text summarization model using clustering techniques. World of Computer Science and Information Technology Jour-nal (WCSIT) (2012)
12. F. Alotaiby, I., Alkharashi, S.F.: Processing large Arabic text corpora: peliminary analysis and results, in *Proceedings of the Second International Conference on Arabic Language Resources and Tools*, pp. 78–82 (2009)
13. M.A. Attia, Arabic tokenization system, in *Proceedings of the 2007 Workshop on Computational Approaches to Semitic Languages: Common Issues and Resources. Association for Computational Linguistics*, pp. 65–72 (2007)
14. I.A. El-Khair, Effects of stop words elimination for Arabic information retrieval: a comparative study. Int. J. Comput. Inf. Sci. **4**(3), 119–133 (2006)
15. T.F. Gharib, M.B. Habib, Z.T. Fayed, Arabic Text Classification Using Support Vector Machines. IJ Comput. Appl. **16**(4), 192–199 (2009)
16. N. Yousef, A. Abu-Errub, A. Odeh, H. Khafajeh, An improved Arabic word's roots extraction method using N-gram technique (2014)
17. A.T. Al-Taani, M.M. Al-Omour, An extractive graph-based Arabic text summarization approach, in *The International Arab Conference on Information Technology*, Jordan (2014)
18. H. Al Ameed, S. Al Ketbi, A. Al Kaabi, K. Al Shebli, N. Al Shamsi, N. Al Nuaimi, S. Al Muhairi, Arabic light stemmer: a new enhanced approach, in *The Second International Conference on Innovations in Information Technology (IIT'05)*, 1–9 (2005)

19. S. Khoja, R. Garside, *Stemming Arabic Text* (Computing Department, Lancaster University, UK, 1999)
20. J.B. Lovins, Development of a stemming algorithm. Mech. Translat. Comp. Linguistics **11**(1–2), 22–31 (1968)
21. A. Chen, F. Gey, Building an Arabic stemmer for information retrieval. TREC **2002**, 631–639 (2002)
22. L.S. Larkey, L. Ballesteros, M.E. Connell, Light stemming for Arabic information retrieval, in Arabic computational morphology (Springer, Dordrecht, 2007), pp. 221–243
23. T. Buckwalter, Issues in Arabic orthography and morphology analysis, in *Proceedings of the Workshop on Computational Approaches to Arabic Script-Based Languages* (Association for Computational Linguistics, 2004), pp. 31–34
24. M. Boudchiche, A. Mazroui, M.O.A.O. Bebah, A. Lakhouaja, A. Boudlal, AlKhalil Morpho Sys 2: A robust Arabic morpho-syntactic analyzer. J. King Saud Univ.-Comput. Inf. Sci. **29**(2), 141–146 (2017)
25. S. Khoja, APT: Arabic part-of-speech tagger, in *Proceedings of the Student Workshop at NAACL* 20–25 (2001)
26. N. Habash, O. Rambow, Arabic tokenization, part-of-speech tagging and morphological disambiguation in one fell swoop, in *Proceedings of the 43rd Annual Meeting of the Association for Computational Linguistics (ACL'05)*, pp. 573–580 (2005)
27. El-Haj, M., Kruschwitz, U., Fox, C.: Using Mechanical turk to create a corpus of arabic summaries, in *Language Resources and Evaluation Conference (LREC)* (May 17–23, Malta, 2010), pp. 36–39

Effects of Light Stemming on Feature Extraction and Selection for Arabic Documents Classification

Yousif A. Alhaj, Mohammed A. A. Al-qaness, Abdelghani Dahou,
Mohamed Abd Elaziz, Dongdong Zhao and Jianwen Xiang

Abstract This chapter aims to study the effects of the light stemming technique on feature extraction where Bag of Words (BoW) and Term frequency- Inverse Documents (TF-IDF) are employed for Arabic document classification. Moreover, feature selection methods such as Chi-square (Chi2), Information gain (IG), and singular value decomposition (SVD) are used to select the most relevant features. K-nearest Neighbor (kNN), Logistic Regression (LR), and Support Vector Machine (SVM) classifiers are used to build the classification model. Experiment are conducted using a public data collected from Arab websites, namely, BBC Arabic dataset. Experiment results show that SVM outperforms LR and KNN. Furthermore, BoW outperforms TF-IDF without using a stemming technique. Using a Robust Arabic Light Stemmer (ARLStem) as our main light stemmer shows a positive effect when combined with TF-IDF over the baseline. In the experiment where Chi2 is used as the feature selection technique, SVM resulted in 0.9568% F1-micro using BoW to extract the features from the dataset where 5000 relevant features were selected. In the experiment where IG is used as the feature selection method, SVM achieved 0.9588% F1-micro with BoW and 4000 selected features. Finally in the experiment where SVD is used as the feature selection technique, SVM reached 0.9569% F1-micro when using BoW and 5000 relevant feature were selected. The aforementioned experiments report the best results achieved where stemming is not employed.

Keywords Arabic text classification · Feature extraction · Feature selection · Stemming techniqueue

Y. A. Alhaj · A. Dahou · D. Zhao · J. Xiang (✉)
Hubei Key Laboratory of Transportation of Internet of Things, School of Computer Science and Technology, Wuhan University of Technology, Wuhan 430070, China
e-mail: jwxiang@whut.edu.cn

M. A. A. Al-qaness
School of Computer Science, Wuhan University, Wuhan 430072, China

M. Abd Elaziz (✉)
Department of Mathematics Faculty of Science, Zagazig University, 44519 Zagazig, Egypt
e-mail: abd_el_aziz_m@yahoo.com

© Springer Nature Switzerland AG 2020
M. Abd Elaziz et al. (eds.), *Recent Advances in NLP: The Case of Arabic Language*, Studies in Computational Intelligence 874,
https://doi.org/10.1007/978-3-030-34614-0_4

1 Introduction

Classifying documents manually is known to be time-consuming and unmanageable because of the huge amount of information gathered every day from different information sources. Thus, it is necessary to introduce an effective automated system for documents classification. Automatic text classification (ATC) is a machine learning approach that focuses on assigning a text document automatically into a thematic category from a predefined set of classes. ATC plays a critical role in a lot of information retrieval applications such as sentiment analysis [1], spam filtering [2], and email filtering [3]. ATC includes many statistical and supervised learning algorithms such as support vector machine (SVM), Naive Bayes (NB), k-nearest neighbors (KNN), decision tree, maximum entropy, hidden Markov model, and neural network. The documents in the ATC pass by different steps [4]. The first step is the pre-processing where document conversion, tokenization, normalization, stop word removal, and stemming are employed. The second step is document modeling which contains feature selection and vector space model construction. The last step is document classification wherein the training and testing data are divided from a dataset. In this step, the classifier builds the classification model from training data and evaluate it on testing data. However, ATC in the Arabic language obtains insufficient performance as compared with other languages like the English language, as Arabic is possess a highly inflectional and derivational nature that makes morphological analysis very complicated.

With the raising of data used in text categorization, the researchers deal with the problem of the high dimensionality of feature space [5]. Many approaches were investigated to solve this problem such as TF-IDF, BoW Feature Extraction approaches, pre-processing tasks such normalization, stop word removal and stemming technique, and feature selection methods which choose the vital features from the data. In general, reducing the dimension of feature space in documents classification could enhance the effectiveness of classifiers, which will reduce the time of processing and the computational process. Nonetheless, optimal feature extraction, selection, and classifier of the vital features help to enhance the accuracy of classification.

Several techniques have been applied to enhance the functionality of the text classification model. Stemming method is an essential technique of pre-processing task which is used to make the classification less dependent, cut down the feature space, enhance the feature representation, and improve the functionality of classification [6]. The stemming is known as the process that reduces the word to their root or stem. Stemming methods are adopted in many applications such as text classification [6], summarization [7], and data compression [8].

The main contribution of this chapter is to study the influence of light stemming on Bag of word (BoW) and Term Frequency-Inverse Document Frequency (TF-IDF) as features extractions approaches, Chi-square (Chi2), information gain and singular value decomposition (SVD) as feature selection methods, and Logistic Regression (LR), KNN, and SVM used as classifiers.

The remaining sections of this chapter are organized as follows; Sect. 2 present the related works. Section 3 presents the research methodology. Experimental series are described in Sect. 4. Finally, the conclusion is presented in Sect. 5.

2 Related Works

This section presents recent works proposed to tackle the classification problem of Arabic documents. Mahmoud et al. [9] have proposed a technique to improve the accuracy of the Arabic text categorization. BoW used as feature extraction. They suggested mixing a bag of words technique with two adjacent words gathered with different proportions. The term frequency was applied as features selection, and the texts were classified using frequency ratio accumulation. Normalization, Information Science Research Institute (ISRI) and Tashaphyne Light Arabic Stemmers were used in this work. To assess this technique, three datasets having different categories were gathered from online Arabic websites. The results showed that the text classification when applying normalization outperformed text classification when normalization and stemming were not adopted in terms of accuracy. The results indicate that using text classification with normalization achieved the highest classification accuracy of 98.61% when using a dataset with 1200 documents distributed among four classes.

Rasha et al. [10] performed a comparative study of the effect of four classifiers on Arabic text categorization accuracy applying stemming method. These classifiers are the Sequential Minimal Optimization (SMO), Naive Bayesian (NB), Decision Tree J48 and KNN. Two stemmers which are Khoja and light stemmers were adopted as preprocessing tasks. The outcome was compared with the baseline of classification without applying stemming. The authors gathered a corpus from local and international newspaper websites. The corpus contains up of 750 documents divided into five classes which are economics, religion, politics, sports, and technology. All documents were pre-processed by removing punctuation marks, digits, the formatting tags, and non-Arabic words. The results were indicated in terms of precision, recall, and F-measure. The findings indicate that light stemmer offered a better accuracy than Khoja stemmer and SMO classifier outperformed the other classifiers in the training stage, while NB classifier outperformed the different classifiers in the test stage. This is since SMO needs massive data to perform better. It achieved 94% in terms of F-measures when using light stemming, and NB classifier is applied. But the authors did not apply feature selection to reduce the dimensionality of the data features.

Harrag et al. [11] investigated the efficiency of classifying Arabic text documents using decision tree. Stop word removal and light stemming used as preprocessing tasks. The functionality of the enhanced classifier reached about 93% of accuracy for the scientific dataset and 91% for the literary dataset. The results indicated that the nature and the specificity of the corpus documents influenced the classifier functionalities.

Baraa et al. [12] proposed a novel text classification technique for Arabic language, which is the frequency ratio accumulation method (FRAM) focusing on features selection and classification. Light stemming, normalization and stop word removal used as preprocessing task. Term Frequency (TF) used as feature extraction. The results showed that the FRAM offered better performance than three classifiers: NB, Multi-variant Bernoulli Naive Bayes (MBNB) and Multinomial Naive Bayes models (MNB). FRAM achieved 95.1% F1-macro by applying the unigram word-level representation approach.

Al-Shargabi et al. [13] studied the influence of stop word removal on Arabic documents classification using several classifiers. They affirmed that the SVM with minimal sequential optimization achieved the highest accuracy and lowest error rate.

AbuZeina and Al-Anzi [14] used the Linear Discriminant Analysis (LDA) as feature reduction for Arabic document classification. They used a dataset that comprises 2000 documents distributed over five classes. Experimental results demonstrated that the performance of LDA is similar to SVD.

Ayedh et al. [4] studied the effect of pre-processing tasks with effects of their combination on Arabic documents classification. The preprocessing tasks including light stemming, normalization and stop word elimination. Three machine learning algorithms which are SVM, KNN and NB are applied to assess the model. TF-IDF applied as feature extraction. They affirmed that SVM outperformed KNN and NB when the normalization and stemming were joined together. Suhad et al. [15] compared the accuracy of the existing Arabic stemming techniques when they deployed the Arabic WorldNet ontology. Several stemming techniques such as Khoja, Light stemmer and Root Extractor. BoW used as feature extraction. Furthermore, they adopted the conceptual representation method as a lexical and semantic technique. They used BBC dataset in their experiments. Normalization, stop word removal and stemming technique used as preprocessing tasks. The authors affirmed that the position tagger with the root extractor offers the most optimal outcomes compared to other stemming techniques. Moreover, the combination of "Has Hyponym" relation with position tagger outperformed other semantic relations and led to an increase of 12.63% compared to different combinations.

Attia et al. [16] designed an alternative framework for Arabic word root extraction and text classification. It is based on the application of Arabic patterns and extracts the root without depending on any dictionary. To assess the functionality, a corpus containing 1526 documents distributed on six categories gathered from the Saudi Press Agency (SAP) was used. During preprocessing tasks, stop words, non-Arabic letters, symbols, and digits were eliminated. BoW was used as feature extraction. LibSVM classifier with three N-gram Kernel (N = 2, 3, 4) was used. They claimed that their results indicate that the root extraction enhances the quality of classifiers in terms of recall, accuracy, F-measures, and the precision was slightly reduced.

Alhaj et al. study the effects of stemming techniques on Arabic document classification [17]. Preprocessing tasks such as normalization, stop word removal and stemming techniques were included. Arabic stemming techniques including Information Science Research Institute (ISRI), ARLStem and Tashaphyne were used to reduce the word into their root or stem. TF-IDF was used as feature extraction and

Chi2 was used as feature selection method. Supervisor learning algorithms such as NB, SVM, and KNN were used to build the classification model. Experiment results indicates that best results achieved with SVM as classifier and ARLStem stemmer.

3 Research Methodology

An Arabic document classification framework typically comprises of three primary phases, the preprocessing phase, feature extraction (representation) phase, and lastly, the document classification phase. The preprocessing phase covers the process which convert the document into an appropriate pattern. Feature extraction phase retrieves the features from the document and changes it into the numerical vector and lastly, document classification phase includes classification model assessment and classification model design. A brief illustration of an Arabic document classification framework is shown in Fig. 1.

The first phase in Fig. 1 is the preprocessing task which involves tokenization, normalization, stop word removal and stemming technique. The second phase includes feature extraction and selection. The third phase start by splitting the dataset into

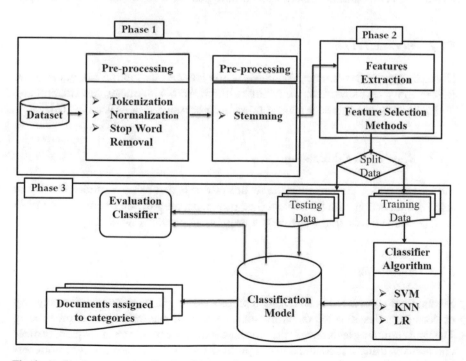

Fig. 1 Arabic document classification framework

training and testing set. Then, the construction and evaluation of the classification model. Framework phases are described in details in the following paragraphs.

3.1 Document Pre-processing

Document preprocessing is the process that transforms the textual words into their accurate format. It is considered as the first step in a document categorization system and used to minimize the vagueness of words to improve the functionality of classification.

3.1.1 Tokenization

Tokenization is considered as the process which divides the documents into tokens (terms). Terms are usually separated from each other by blanks by (commas, whitespace, quotes, semicolons, and periods). These tokens could be single words (verb, noun, conjunction, pronoun, preposition, numbers, article, punctuation, and alphanumeric) which transforms without concentrating on their relationships or meanings.

3.1.2 Normalization

Normalization is the manner that changes the letter in text into a canonical form. Before apply normalization, we remove all-non Arabic character, punctuation and digits. We used the same normalization as preprocessing task as used in [18].

3.1.3 Stop Word Elimination

Stop words which are known as the recurrent words that bring no understanding or signs in connection with the content (i.e., pronouns, prepositions, conjunctions). We used the same list of stop words which used in our previous work [19].

3.1.4 Stemming

Stemming is considered as an essential preprocessing phase before tackling any task of document classification or information retrieval in natural language processing. The light stemming technique is the process of stripping off the most frequent suffixes and prefixes using a predefined list of prefixes and suffixes. Several light stemmers have been suggested for the Arabic language such in [20]. Nevertheless, light stemming has many challenges in removing prefixes or suffixes in Arabic. In some cases, the elimination of a fixed set of prefixes and suffixes without checking if the leftover word is stem can lead to an unexpected outcome, especially when distinguishing other letters from root letters. In our study we used ARLStem light stemmer [21] according to its performance in [17], which compared the functionality of light and root base stemmers and affirmed that ARLStem light stemmer significantly outperforms the root base stemmer.

3.2 Document Modeling

This technique can be termed as feature extraction or document indexing that includes two phases which are feature extraction (Feature Representation), and feature selection which are described in details in the following paragraphs.

3.2.1 Feature Extraction

The process that transforms the text document into a numeric vector is known as feature extraction or feature representation. Several feature extraction techniques are applied to extract the features from texts and translated into vectors such as BoW, and TF-IDF [22]. BoW extract features by firstly count all the unique words from the documents as shown in Fig. 2. Then score the words in each document. In BoW the order of words within documents is ignored, disregarding grammar and their occurrence frequencies are adopted [23]. In BoW there are a lots of zeros and it can be called sparse representation or a sparse vector. Sparse vectors demand more computational resources and memory that will increase the dimensions and make the process very challenging.

TF-IDF it is a statistical measure used to evaluate the importance of a word in a given document as shown Fig. 3. Term Frequency (TF) as well known bag of words and Inverse Document Frequency (IDF) is known as the method to be applied in conjunction with TF in order to decrease the impact of implicitly common terms in the documents. TF-IDF mathematical extraction of the weight of a word in a document is given in Eq. 1.

$$W(d, t) = TF(d, t) * log(\frac{N}{df(t)}) \tag{1}$$

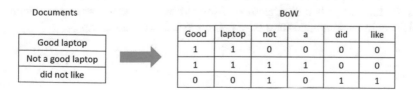

Fig. 2 BoW Feature extraction

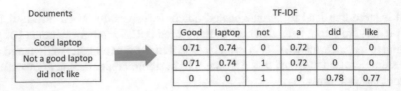

Fig. 3 TF-IDF feature extraction

where N represent to the number of documents, and $df(t)$ represent to the number of documents comprising the term t in the documents or corpus.

3.2.2 Chi-Square Testing χ_2

χ_2 is defined as a well-known discrete data hypothesis testing approach from statistics. This technique assesses the correlation between two variables and determines whether these variables are independent or correlated [24]. (χ_2) Value for each term t in a category c can be defined by applying Eqs. 2, and 3.

$$\chi_2(t_k, c_i) = \frac{|Tr|.[p(t_k, c_i) * P(t_k^-, c_i^-) - p(t_k^-, c_i^-) * p(t_k^-, c_i)]^2}{p(t_k) * p(t_k^-) * p(c_i^-)} \qquad (2)$$

Moreover, is evaluated using

$$\chi_2(t, c_i) = \frac{N * (AD - CB)^2}{(A + C) * (B + D) * (C + D)} \qquad (3)$$

where A denotes the number of documents in category c including the term t. B is the number of documents not in category c including the term t. C is the number of documents in category c not including the term t, D is the number of documents of a different category not comprising term t, and N is the amount number of documents.

3.2.3 Information Gain

IG is an significant feature selection technique which evaluates how much the attributes is informative about the category. IG present the doubt reduction in naming class by knowing when the measure of the attribute. IG is a rating score technique which could be computed for a term using the following equation.

$$IG(w) = -\sum_{j=1}^{k} P(C_j)log(P(C_j)) + P(w)$$

$$+P(w)\sum_{j=1}^{k} P(C_j|w)log(P(C_j|w)) + P(\bar{w})$$

$$+P(\bar{w})\sum_{j=1}^{k}(P(C_j|\bar{w})log(P(C_j|\bar{w}))) \qquad (4)$$

where C is a category dimension and can be K categories, it is announced as $\{C_1,...,C_k\}$. The probability of a feature $P(C_j)$ is the divide by the number of documents that related to category C_j out of amount documents, and $P(w)$ is the divide of documents in which word w passes. $P(C_j|w)$ is calculated as the divide of documents from category C_j that has word W. Where $\bar{w}$ intends that a document does not comprise the word w.

3.2.4 Singular Value Decomposition (SVD)

In field of information retrieval, SVD method is suggested by [25] for reducing the attributes in text classification where documents are symbolized with vectors. SVD method can decompose a matrix Q of the size $m \times n$ into the production of three matrices: a $m \times m$ perpendicular matrix u, an $m \times n$ diametrical matrix B, and the refer of an perpendicular matrix K of the size $n \times n$. SVD formulation is usually demonstrated as follows:

$$Q_{n \times m} = R_{m \times m} \times B_{m \times n} \times K_{n \times n}^{T} \qquad (5)$$

Attributes reduction is used to remove irrelevant data from rows from the underside of matrices R and B. As well, getting rid of left columns from matrices B and K^T.

3.3 Document Classification

In this step, three machine learning algorithms are studied which are commonly accessible in machine learning algorithm and statistical classification. Machine learning algorithms such as LR [26], KNN [27], and SVM [28] are used to assess the influence of stemming approach on BoW and TF-IDF feature extractions for Arabic document classification. In KNN, the number of neighbors is set to 6, the LR classifier the random_state is set to false, and SVM is implemented with a linear kernel where C = 1.

4   Experiment Work

In this section, an in-depth investigation was carried out to measure the performance of term feature representation methods, feature selection and classification algorithms in a range of feature size grounded on stemming. The effect of stemming on feature representations and feature selection on the accuracy of Arabic documents classification can be inferred according to the experimental results. In the following subsections, the adopted datasets and accuracy analysis are briefly described.

4.1   Dataset

We evaluated the effect of a light stemmer on different feature extractions and selection on a well-known publicly available Arabic dataset namely, BBC Arabic dataset which was obtained by Saad et al. [29]. The statistics and the data class distribution are described in Table 1 and Fig. 4.

4.2   Accuracy Analysis

With regards to these experiments, all documents in the dataset were arranged by converting them into UTF-8 encoding. We begin the preprocessing by eliminating all non-Arabic punctuation, characters, and digits. After that, normalization has been applied to normalize each character to its standard form. A list of stop words were eliminated from the dataset. Then TF-IDF and BoW were applied to retrieve the features from text documents. Feature selection approach is used to reduce the dataset by selecting important features where 1000, 2000, 3000, 4000, and 5000 are the selected features sets. LR, KNN, and SVM classifiers were used to assess the precision of the document classification system. Cross-validation was carried out in every

Table 1   The statistic of BCC dataset

Category name	Number of documents
Middle east news	2356
Business and economy	296
Misc	122
Science and technology	232
Sport	219
World News	1489
Newspapers highlights	49
Total	4763

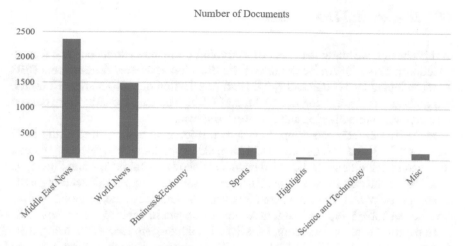

Fig. 4 Samples distributed in BCC dataset

classification experiments that separated data into ten mutually exclusive subsets known as folds. Every fold nearly comprises 476 documents. One of the subsets is adopted as a test set, where the remaining subsets are considered as training sets. Many mathematic rules like recall (R), precision (P), and F-measure (F) are studied to assess the functionality of classification model to categorize the documents into the correct category which are defined as follows.

$$Precision = \frac{TP}{TP + FP} \tag{6}$$

$$Recall = \frac{TP}{TP + FN} \tag{7}$$

where TP is used to denote the number of documents that are correctly assigned to the category. TN stands for indicating the number of documents that are correctly assigned to the negative group. FP is the number of documents a system wrongly attributed to the class. FN symbolizes the number of documents that fall into the class but are not assigned to the category. F1-Micro is calculated as follows.

$$F1 - Micro = \frac{2 \times Precision \times Recall}{Precision + Recall} \tag{8}$$

4.3   Results and Discussions

In this chapter, we study the effect of stemming on feature extractions and feature selection methods for Arabic Document classification. Extracted features from BBC dataset adopted by TF-IDF and BoW. Feature selection methods namely Chi2, IG, and SVD are used alongside LR, SVM, and KNN. The baseline (BS) is the dataset after applying normalization and stop word removal.

In the first experiments setup, F1-Micro results of stemming effect using BBC dataset on BoW and TF-IDF where Chi2 is used as default feature selection method are listed in Table 2 and the best fold results are shown in Tables 3, 4 and Figs. 5, 6.

In the second experiments setup, F1-Micro results of stemming effect using BBC dataset on BoW and TF-IDF where IG is used as default feature selection method are listed in Table 5 and the best fold results are shown in Tables 6, 7 and Figs. 7, 8.

In the third experiments setup, F1-Micro results of stemming effect using BBC dataset on BoW and TF-IDF where SVD is used as default feature selection method are listed in Table 8 and the best fold results are shown in Tables 9, 10 and Figs. 9, 10.

According to the conducted study, applying a stemming technique has a disparate impact on the accuracy of Arabic documents classification based on feature extractions, feature selection and classifiers.

Concerning first experiments setup, Table 2 results indicate that combining BoW feature representation with stemming has a negative impact. The results are precise where the baseline shows better performance compared to applying the stemming

Table 2 F1-Micro measure scores on BoW and TF-IDF and Chi2

Classifier	Feature size	BoW		TF-IDF	
		BS	Stemmer	BS	Stemmer
LR	1000	0.8467	0.8457	0.777	0.8289
	2000	0.8585	0.8585	0.8224	0.8579
	3000	0.8671	0.8612	0.8438	0.8749
	4000	0.8738	0.8633	0.857	0.8774
	5000	0.8757	0.8681	0.8681	0.8814
KNN	1000	0.8171	0.8084	0.7806	0.8064
	2000	0.8018	0.7921	0.7859	0.7934
	3000	0.7892	0.7772	0.7626	0.7972
	4000	0.784	0.7743	0.7642	0.7899
	5000	0.7827	0.7651	0.7473	0.7945
SVM	1000	0.9341	0.9169	0.8006	0.8665
	2000	0.9542	0.9326	0.8587	0.8791
	3000	0.9475	0.9295	0.8694	0.8845
	4000	0.9565	0.9315	0.8696	0.8872
	5000	0.9568	0.9334	0.8705	0.8896

Table 3 F1-Micro by class for SVM on BoW and Chi2 in the best fold

Class	Precision	Recall	F1-Micro
Middle east news	0.971	0.975	0.973
World News	0.953	0.960	0.957
Business and economy	0.968	1.000	0.984
Sports	0.955	0.955	0.955
Highlights	1.000	1.000	1.000
Science and technology	1.000	0.957	0.978
Misc	1.000	0.833	0.909
W.Avg	0.967	0.967	0.966

Table 4 F1-Micro by class for SVM on TF-IDF and Chi2 in the best fold

Class	Precision	Recall	F1-Micro
Middle east news	0.908	0.966	0.936
World news	0.898	0.886	0.892
Business and economy	0.920	0.767	0.836
Sports	0.952	0.909	0.930
Highlights	1.000	1.000	1.000
Science and technology	0.857	0.783	0.818
Misc	1.000	0.583	0.737
W.Avg	0.909	0.908	0.906

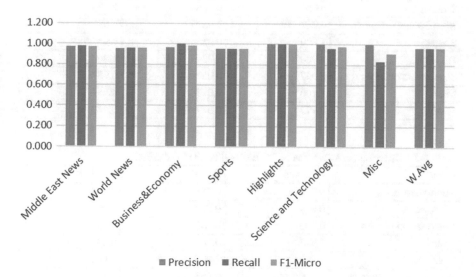

■ Precision   ■ Recall   ■ F1-Micro

Fig. 5 F1-Micro by class for SVM on BoW and Chi2 in the best fold

Table 5 F1-Micro measure scores on BoW, TF-IDF and IG

Classifier	Feature size	BoW		TF-IDF	
		BS	Stemmer	BS	Stemmer
LR	1000	0.9517	0.9343	0.7811	0.8148
	2000	0.9507	0.9364	0.8117	0.8467
	3000	0.9515	0.9381	0.8237	0.8545
	4000	0.9507	0.9412	0.83	0.8551
	5000	0.9513	0.9412	0.8354	0.8583
KNN	1000	0.8031	0.8039	0.7533	0.7863
	2000	0.7563	0.7818	0.7567	0.8077
	3000	0.7409	0.76	0.7649	0.8061
	4000	0.7235	0.7583	0.7571	0.798
	5000	0.7004	0.7512	0.7594	0.7896
SVM	1000	0.9484	0.925	0.8184	0.8511
	2000	0.9565	0.9351	0.8576	0.881
	3000	0.9567	0.9347	0.8707	0.8837
	4000	0.9588	0.9393	0.8742	0.8858
	5000	0.958	0.9425	0.8753	0.8854

Table 6 F1-Micro by class for SVM on BoW and IG in the best fold

Class	Precision	Recall	F1-Micro
Middle east news	0.975	0.992	0.983
World news	0.980	0.966	0.973
Business and economy	0.967	0.967	0.967
Sports	1.000	0.955	0.977
Highlights	1.000	1.000	1.000
Science and technology	0.957	0.957	0.957
Misc	1.000	0.917	0.957
W.Avg	0.977	0.977	0.977

Table 7 F1-Micro by class for SVM on TF-IDF and IG in the best fold

Class	Precision	Recall	F1-Micro
Middle east news	0.901	0.966	0.933
World news	0.903	0.879	0.891
Business and economy	0.920	0.767	0.836
Sports	0.955	0.767	0.836
Highlights	1.000	1.000	1.000
Science and technology	0.900	0.783	0.837
Misc	1.000	0.583	0.737
W.Avg	0.909	0.908	0.906

Effects of Light Stemming on Feature Extraction …

Table 8 F1-Micro measure scores on BoW and TF-IDF and SVD

Classifier	Feature size	BoW		TF-IDF	
		BS	Stemmer	BS	Stemmer
LR	1000	0.9397	0.9282	0.8543	0.8616
	2000	0.9473	0.9391	0.8547	0.8621
	3000	0.9498	0.9425	0.8512	0.8614
	4000	0.9505	0.9429	0.8505	0.8614
	5000	0.9505	0.9427	0.8501	0.8614
KNN	1000	0.7928	0.7863	0.8455	0.8543
	2000	0.7405	0.7596	0.8406	0.8453
	3000	0.7031	0.7451	0.8331	0.8402
	4000	0.6721	0.7298	0.8373	0.8419
	5000	0.6698	0.7321	0.8408	0.8417
SVM	1000	0.9374	0.9238	0.886	0.8885
	2000	0.9511	0.9345	0.8868	0.8912
	3000	0.954	0.9404	0.8935	0.8931
	4000	0.9553	0.9444	0.8956	0.8931
	5000	0.9569	0.9452	0.8954	0.8931

Table 9 F1-Micro by class for SVM on BoW and SVD in the best fold

Class	Precision	Recall	F1-Micro
Middle east news	0.971	0.983	0.977
World news	0.967	0.973	0.970
Business and economy	0.931	0.900	0.915
Sports	0.955	0.955	0.955
Highlights	1.000	1.000	1.000
Science and technology	0.955	0.913	0.933
Misc	1.000	0.833	0.909
W.Avg	0.967	0.967	0.966

Table 10 F1-Micro by class for SVM on TF-IDF and SVD in the best fold with best fold

Class	Precision	Recall	F1-Micro
Middle east news	0.905	0.970	0.937
World news	0.907	0.852	0.879
Business and economy	0.920	0.767	0.836
Sports	0.957	1.000	0.978
Highlights	1.000	0.800	0.889
Science and technology	0.818	0.783	0.800
Misc	1.000	0.833	0.909
W.Avg	0.908	0.908	0.906

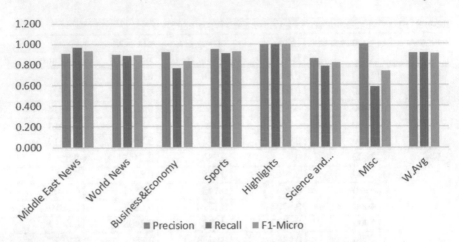

Fig. 6 F1-Micro by class for SVM on TF-IDF and Chi2 in the best fold

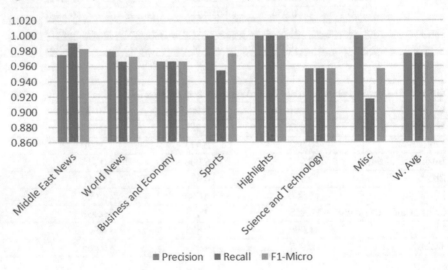

Fig. 7 F1-Micro by class for SVM on BoW and IG in the best fold

technique with all feature size and all classifiers. Moreover, the best results achieved when using BoW with SVM with the number of selected features is equal to 5000 using Chi2. From Fig. 6 and Table 3 the highest performance of precision, recall, and F1-Micro achieved on Highlights category and the lowest performance of precision, recall, and F1-Micro achieved on sports and Misc categories, respectively. Moreover, using TF-IDF and applying stemming alongside Chi2 shows a positive impact. The stemmer results improve the accuracy with all feature size and all classifiers. As well, the best results achieved using TF-IDF and SVM when the number of selected features is equal to 5000 using Chi2. From Fig. 4 and Table 4 the highest performance

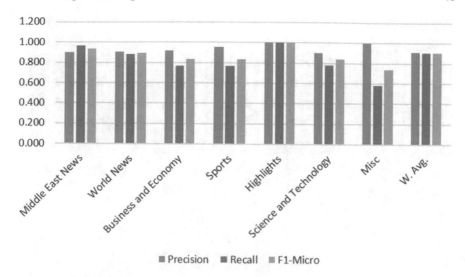

Fig. 8 F1-Micro by Class for SVM on TF-IDF and IG in the best fold

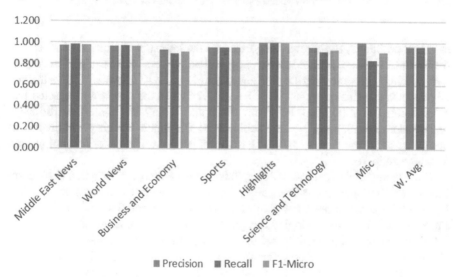

Fig. 9 F1-Micro by class for SVM on BoW and SVD with best fold

of precision, recall, and F1-Micro achieved on Highlights category and the lowest performance of precision, recall and F1-Micro achieved on Science, Technology and Misc categories, respectively.

Concerning second experiments setup, Table 5 results also indicates that with BoW combined with stemming shows a negative impact. The highest results achieved using BoW and SVM when the number of selected features is equal to 4000 using IG. From Fig. 7 and Table 6 the highest performance of precision, Recall, and F1-Micro

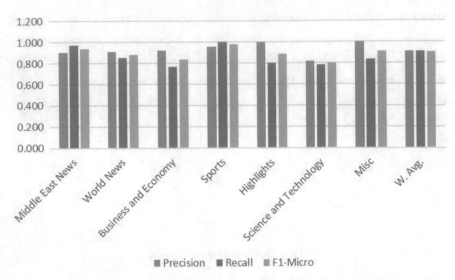

Fig. 10 F1-Micro by Class for SVM on TF-IDF and SVD in the best fold

achieved on Sports, Highlights categories and the lowest performance of precision, recall, and F1-Micro achieved on Science, technology, and Misc categories, respectively. Moreover, combining TF-IDF and stemming with IG shows a positive impact. Highest results achieved using TF-IDF and SVM when the number of selected features is equal to 4000 using IG. From Fig. 8 and Table 7 the highest performance of Precision, Recall, and F1-Micro achieved on Highlights category and the lowest performance of Precision, Recall and F1-Micro founded on Science, Technology and Misc categorie, respectively.

Concerning third experiments setup, Table 8 results also indicates that BoW combined with stemming shows a negative impact. The highest results achieved using BoW and SVM when the number of selected features is equal to 5000 using SVD. From Fig. 9 and Table 9 the highest performance of precision, recall, and F-Measure achieved on Highlights category and the lowest performance of precision, recall, and F1-Micro achieved on Business, Economy, Misc categories, respectively. Moreover, combining TF-IDF with stemming and SVD shows a negative impact. Highest results achieved using TF-IDF and SVM when the number of selected features is equal to 4000 using SVD. From Fig. 10 and Table 10 the highest performance of precision, recall, and F1-Micro achieved on Highlights and Sport categories and the lowest performance of Precision, Recall and F1-Micro founded on Science, Technology, Business, and Economy categories, respectively.

It has been discovered that the accuracy of the classifier has uncertain outcome depending on the type of feature representation techniques, classifier algorithms, and the number of features selected by feature selection technique from the dataset. Applying stemming without focusing on the type of feature representation, classifier, and the number of features may lead to a negative impact. Nevertheless, the researchers should take into consideration when they are dealing with stemming on Arabic text classification the selection of the optimal feature extraction, feature selection, and classifier algorithms. This study provides information in order to determine the appropriate method of feature representation, classification algorithm, features size, and stemming technique without applying them directly.

5 Conclusion

In this chapter, the investigators examine the influence of the light stemming technique on BoW, TF-IDF feature extraction and feature selection methods on the accuracy of Arabic document classification. Several classifiers have been considered to solve the problem of text classification such as LR, KNN, SVM. Chi2, IG, SVD feature selection methods used to remove the unnecessary features and select the essential features. Results of this study indicate that the applied stemmer affords a slight enhancement and assistance to progress the presentation, accuracy on Arabic text classification by considering the feature extraction method, machine learning classifier, and feature selection method. The outcomes noticeably show the superiority of the SVM over the LR and KNN classifiers using TF-IDF and BoW. The best results achieved combining TF-IDF with stemming while good result achieved using BoW representation without stemming. The study conclusions assistance the auditors on Arabic document classification to use the stemming technique with ideal feature extraction, machine learning classifier, feature selection method with important features without applying it directly. Experimental results will be helpful for recherches in Arabic document classification in choosing optimal feature extractions, feature selection, features size and optimal classifier when they are dealing with light stemming such as ARLStem stemmer.

Future research will focus on the study of the effects of stemming technique in different feature extraction such as word embedding, feature selection methods and deep learning.

Acknowledgements This work was partially supported by the National Natural Science Foundation of China (Grant No. 61672398, 61806151), the Defense Industrial Technology Development Program (Grant No. JCKY2018110C165), and the Hubei Provincial Natural Science Foundation of China (Grant No. 2017CFA012).

References

1. A. Dahou, M.A. Elaziz, J. Zhou, S. Xiong, Arabic sentiment classification using convolutional neural network and differential evolution algorithm. Comput. Intell. Neurosci. **2019** (2019)
2. J.R. Méndez, T.R. Cotos-Yañez, D. Ruano-Ordás, A new semantic-based feature selection method for spam filtering. Appl. Soft Comput. **76**, 89–104 (2019)

3. S. Sakurai, A. Suyama, An e-mail analysis method based on text mining techniques. Appl. Soft Comput. **6**(1), 62–71 (2005)
4. A. Ayedh, G. Tan, K. Alwesabi, H. Rajeh, The effect of preprocessing on arabic document categorization. Algorithms **9**(2), 27 (2016)
5. J.-S. Kuo, Active Learning for Constructing Transliteration. J. Am. Soc. Inf. Sci., **59**(1), 126–135 (2008). [Online]. Available: http://ejournals.ebsco.com/direct.asp?ArticleID=40729F4826A638E14483
6. A. Ayedh, G. Tan, Building and benchmarking novel Arabic stemmer for document classification. J. Comput. Theor. Nanosci. **13**(3), 1527–1535 (2016)
7. Slamet, C., Atmadja, A.R., Maylawati, D.S., Lestari, R.S., Darmalaksana, W., Ramdhani, M.A.: Automated text summarization for indonesian article using vector space model. IOP Conf. Ser. Mater. Sci. Eng. **288**(1) (2018)
8. A. Sinaga, Adiwijaya, H. Nugroho, Development of word-based text compression algorithm for Indonesian language document, in 2015 3rd International Conference on Advanced Information and Communication Technology ICoICT 2015, pp. 450–454 (2015)
9. M. Hussein, H.M. Mousa, R.M. Sallam, Arabic text categorization using mixed words. I.J. Inf. Technol. Comput. Sci. Inf. Technol. Comput. Sci., **11**(11), 74–81, 2016. [Online]. Available: http://www.mecs-press.net/ijitcs/ijitcs-v8-n11/IJITCS-V8-N11-9.pdf
10. R. Mamoun, M. Ahmed, Arabic text stemming: Comparative analysis. in *Conference of Basic Sciences and Engineering Studies (SGCAC)*. IEEE **2016**, 88–93 (2016)
11. F. Harrag, E. El-qawasmeh, I. Al, Improving arabic text categorization using decision trees. First Int. Conf. Networked Digit. Technol. 2009. NDT '09. no. September, pp. 110–115 (2009)
12. B. Sharef, N. Omar, Z. Sharef, An Automated arabic text categorization based on the frequency Ratio Accumulation **11**(2), 213–221 (2014)
13. B. Al-Shargabi, F. Olayah, W.A. Romimah, An experimental study for the effect of stop words elimination for arabic text classification algorithms. Int. J. Inf. Technol. Web Eng. (IJITWE) **6**(2), 68–75 (2011)
14. D. AbuZeina, F. Al-Anzi, Employing fisher discriminant analysis for Arabic text classification. Comput. Electr. Eng. 1–13 (2017)
15. S.A. Yousif, V.W. Samawi, I. Elkabani, R. Zantout, The effect of combining different semantic relations on arabic text classification. World Comput. Sci. Inform. Technol. J **5**(1), 12–118 (2015)
16. A. Nehar, D. Ziadi, and H. Cherroun, "Rational kernels for Arabic Root Extraction and Text Classification," *J. King Saud Univ. - Comput. Inf. Sci.*, vol. 28, no. 2, pp. 157–169, 2016. [Online]. Available: http://dx.doi.org/10.1016/j.jksuci.2015.11.004
17. Y. A. Alhaj, J. Xiang, D. Zhao, M. A. Al-Qaness, M. A. Elaziz, and A. Dahou, "A study of the effects of stemming strategies on arabic document classification," *IEEE Access* (2019)
18. L.S. Larkey, L. Ballesteros, M.E. Connell, Improving stemming for Arabic information retrieval, in *Proceedings of the 26th annual international ACM SIGIR conference on research and development in information retrieval, SIGIR '03*, 2002, p. 275. [Online]. Available: http://portal.acm.org/citation.cfm?doid=564376.564425
19. Y.A. Alhaj, W.U. Wickramaarachchi, A. Hussain, M.A. Al-Qaness, and H.M. Abdelaal, Efficient feature representation based on the effect of words frequency for arabic documents classification, in *Proceedings of the 2nd International Conference on Telecommunications and Communication Engineering* (ACM, 2018), pp. 397–401
20. L. Larkey, L. Ballesteros, and M. Connell, Light stemming for Arabic information retrieval," *Arab. Comput. Morphol.*, pp. 221–243 (2007)
21. K. Abainia, S. Ouamour, H. Sayoud, A novel robust arabic light stemmer. J. Exp. & Theor. Artif. Intell. **29**(3), 557–573 (2017)
22. K. Kowsari, K. Jafari Meimandi, M. Heidarysafa, S. Mendu, L. Barnes, and D. Brown, Text classification algorithms: a survey. Information, **10**(4), 150 (2019)
23. A.K. Uysal, S. Günal, S. Ergin, E.Ş. Günal, Detection of sms spam messages on mobile phones, in *20th Signal Processing and Communications Applications Conference (SIU)*. IEEE **2012**, 1–4 (2012)

24. F. Thabtah, M. Eljinini, M. Zamzeer, W. Hadi, *"Naïve bayesian based on chi square to categorize arabic data,"* in proceedings of The 11th International Business Information Management Association Conference (IBIMA) Conference on Innovation and Knowledge Management in Twin Track Economies (Egypt, Cairo, 2009), pp. 4–6
25. G.W. Furnas, S. Deerwester, S.T. Dumais, T.K. Landauer, R.A. Harshman, L.A. Streeter, K.E. Lochbaum,: Information retrieval using a singular value decomposition model of latent semantic structure," in *11th Annu. Int. ACM SIGIR Conf. Res. Dev. Inf. Retr. (SIGIR 1988)* (1988)
26. P. Tsangaratos, I. Ilia, Comparison of a logistic regression and naïve bayes classifier in landslide susceptibility assessments: The influence of models complexity and training dataset size. Catena **145**, 164–179 (2016)
27. M. Syiam, Z.T. Fayed, M.B. Habib, An intelligent system for arabic text categorization. Int. J. Intell. Comput. Inf. Sci. **6**(1), 1–19 (2006)
28. A. Moh, A. Mesleh, Chi Square Feature Extraction Based Svms Arabic Language Text Categorization System. J. Comput. Sci. **3**(6), 430–435 (2007)
29. M. Saad, W. Ashour, OSAC: Open Source Arabic Corpora, in 6th international conference on computer systems (EECS'10), Nov 25-26, 2010, Lefke, Cyprus., pp. 118–123, 2010. [Online]. Available: http://site.iugaza.edu.ps/msaad/files/2010/12/mksaad-OSAC-Open-Source-Arabic-Corpora-EECS10-rev8.pdf

Improving Arabic Lemmatization Through a Lemmas Database and a Machine-Learning Technique

Driss Namly, Karim Bouzoubaa, Abdelhamid El Jihad
and Si Lhoussain Aouragh

Abstract Lemmatization is a key preprocessing step and an important component for many natural language applications. For Arabic language, lemmatization is a complex task due to Arabic morphology richness. In this paper, we present a new lemmatizer that combines a lexicon-based approach with a machine-learning-based approach to get the lemma solution. The lexicon-based step provides a context-free lemmatization and the most appropriate lemma according to the sentence context is detected using the Hidden Markov Model. The developed lemmatizer evaluations yield to over than 91% of accuracy. This achievement outperforms the state of the art Arabic lemmatizers.

Keywords Arabic NLP · Arabic lemmatization · Lexicon-based lemmatization · Machine-learning-based lemmatization · Hidden markov model · Viterbi algorithm

D. Namly (✉) · K. Bouzoubaa
Mohammadia School of Engineers, Mohammed V University, Rabat, Morocco
e-mail: namly_driss@yahoo.fr

K. Bouzoubaa
e-mail: karim.bouzoubaa@emi.ac.ma

A. El Jihad
Institute of Arabization Studies and Research, Mohammed V University, Rabat, Morocco
e-mail: jihad.hamid@gmail.com

S. L. Aouragh
Faculty of Legal, Economic and Social Sciences - Sale, Mohammed V University, Rabat, Morocco
e-mail: aouragh@hotmail.com

© Springer Nature Switzerland AG 2020　　　　　　　　　　　　　　81
M. Abd Elaziz et al. (eds.), *Recent Advances in NLP: The Case of Arabic
Language*, Studies in Computational Intelligence 874,
https://doi.org/10.1007/978-3-030-34614-0_5

1 Introduction

With more than 414 million native speakers [1], Arabic is the official language of 26 states, one of the six official languages in the United Nations, the fifth most widely spoken language[1] and the most widely used semitic language [2]. Therefore, in the high technologies' era, this magnitude in human capital is reflected on the Internet both in terms of user's number and web content. According to the 2018 World Bank statistics [1], in the Arab world, there are 219 million Internet users and an average annual increase of 66.24% in the number of secure Internet servers between 2010 and 2017.

This rapid enlargement in Arabic web content generates a high demand for relevant applications and services. And since Natural Language Processing (NLP) systems effectiveness is directly related to data space dimension, a normalization technique is required before processing text to reduce data dimension. Lemmatization is a productive way to do that as reflected in Arabic Natural Language Processing (ANLP) applications such as machine translation [3], document clustering [4] or text summarization [5].

Arabic lemmatization is the process of reducing a word to its canonical representation, base form, dictionary form or lexical entry, called the lemma of the word. The Arabic lemma is the word before undergoing any inflection. The lemma for verbs is the perfective third person masculine singular form without clitics. For nouns and adjectives, the lemma corresponds to the nominative singular masculine (or feminine if the word does not accept the masculine) form without clitics, and the particle without clitics for particles. For example, the lemmas of the words "فَكَتَبْتُهَا، المَكْتَبَاتُ، وعَلَيْكُمَا"(and I wrote it, the libraries, and up on you) are respectively "كَتَبَ، مَكْتَبَةٌ، عَلَى"(write, library, on).

When it is not diacritized and without context, Arabic lemmatization does not remove the ambiguity related to the correct lemma among the potential ones. For instance, the word "ولم"(wlm) admits a set of lemmas such as "أَلِمَ"(be insane), "أُمَّة"(group), "لَمَّ"(gather), "لَمْ"(not), "لِمَ"(why), "وَلَمْ"(Belt), "وَلَمَ"(to eat in a feast), whereas in the sentence "وعده ولم يف بوعده"(he promised and did not fulfill it), the word "ولم"(whm) admits a single lemma which is "لَمْ"(not). Thereby, the sentence context is used to remove the ambiguity related to the appropriated lemma.

Although recent studies [3, 5–7] show that lemmatization is the suitable way to enhance the performances and the efficiency of many ANLP applications, very often NLP systems make use of root-based or stem-based stemming [8–10] to cluster words derived from the same stem or root. From an efficiency point of view, relying on a root-based stemming, a NLP algorithm may yield both relevant and not relevant information in their process. For example, the words "كَاتِبُونَ كَتَبْتُ، كِتَابُ، مَكْتَبٌ"(I wrote, a book, an office, writers) have the same root "كتب"(ktb), so clustering these words by root gives a lot of inappropriate and semantically far results. Also, stem-based

---

[1] https://www.babbel.com/en/magazine/the-10-most-spoken-languages-in-the-world   Retrieved March 28, 2018.

clustering gives bad results, because stemming may detect neither words semantic similarity nor syntactic similarity. For example, the stemming result for the words "كَتَبْتُ، يَكْتُبُونَ"(I wrote, they write) gives different stems although the words are inflected from the same verb which is "كَتَبَ"(to write). Moreover, stemming provides two different stems for the words "كَاتِبٌ"(writer) and "كُتَّابٌ"(writers) even if the first is the singular of the second. From a performances' point of view, lemmatization reduces indexing data dimension more than stemming. The reason is that, stemming removes clitics from words, whereas lemmatization removes clitics and inflectional affixes to return the dictionary form of the word. For instance, the verb "كَتَبَ"(to write) supply more than 70 stems such as "أَكْتُبُ"(I write), "كَتَبْتُمَا"(you wrote) or "يَكْتُبُونَ"(they are writing), and one lemma. Thus, instead of indexing the verb by 70 entries, we index it with a single one.

On the other hand, lemmatization is not sufficiently addressed, and efforts dealing with this issue undergo three main drawbacks. Firstly, the lemmatization function is integrated within the morphological analyzers [11–13]. i.e., to lemmatize a text, it is necessary to analyze it with the morphological analyzer and extract the lemma property form the output. This practice makes the lemmatization exploitation difficult for a non-specialist and requires more time. Secondly, some available lemmatizers provide an undiacritized outputs [14]. The lack of diacritics leads to significant lexical ambiguity. For instance, returning "كتب"(ktb) as a lemma is inappropriate since that word has 17 possible diacritizations. Thirdly, stemming and lemmatization are sometimes used interchangeably [15, 16].

The present work proposes a solution to overcome the previous shortcomings. For that purpose, we offer a lemmatization tool, without relying on a morphological analyzer, and that returns all acceptable diacritized for each word in the input sentences. The first lemma of the output list is the most appropriate one according to the sentence context.

In the rest of the paper, related works are reviewed in Sect. 2. In Sect. 3, we explain the proposed approach. Section 4 exposes the evaluation and the comparison with existing works, furthermore, we analyze the effect of the proposed lemmatizer on a text classification task. Conclusions are presented in Sect. 5.

2 State of the Art

Existing works demonstrate that there are mainly three approaches to Arabic lemmatization: lemmatization through morphological analysis, statistical lemmatization and hybrid techniques.

2.1   Morphological Analysis Approach

Some of the handful Arabic morphological analyzers are available while others are whether commercial applications or published but not-available. Among those known in the literature we identify Xerox Arabic Morphological Analysis [17], Buckwalter Arabic Morphological Analyzer [18], ElixirFM [19], Qutuf [20], SALMA [21], Alkhalil morpho sys [13, 22], MADAMIRA [11] and CALIMA-star [12]. We limit this morphological analyzer's review to freely available ones offering the lemma tag.

Standard Arabic Language Morphological Analysis (SALMA) is a morphological analyzer proposed by Sawalha et al. [21]. It consists of several modules which can be used independently to perform a specific task such as root extraction, lemmatization and pattern extraction. They can also be used together to produce the full detailed analysis of a word. We note that SALMA download link is unavailable.

MADAMIRA [11] produces a contextually ranked list of morphological analyses for an input text. It is the outcome of the mix of two tools: MADA [23] and AMIRA [24]. It makes use of SAMA analyzer [25] to get a free of context words analysis, then applies the Support Vector Machine and N-gram language models techniques on an annotated corpus to rank the analyses. It combines a number of tools such as word segmentation, POS Tagging, stemming and lemmatization.

Alkhalil morpho sys II (Alkhalil 2) [13] is the enhanced version of Alkhalil morpho sys, the open source morphological analyzer. Alkhalil 2 provides all possible solutions for the input text out of context. It delivers a set of morpho-syntactic features namely the vowelized form of the word, the root, the stem, the stem pattern, the clitics, the POS tag, the lemma, the lemma pattern and the syntactic state (case for nouns and mood for verbs).

CALIMA-star [12] is an out-of-context morphological analyzer and generator implemented in Python, providing morphological features such as tokenization, phono-logical representation, root, pattern, POS, lemma, gender, number, state (definite-ness), case or lexical rationality (+Human).

From the above, the lemmatization through morphological analysis carry an out of context output for all analyzers except for MADAMIRA.

2.2   Statistical Approach

Attia et al. [26] attempt to relate surface word forms to their lemmas using a machine learning classifier (Decision Trees (C4.5) and Extra Trees). They map the stem to the pattern of the lemma; then, they map the pattern of the lemma to the lemma form, by extracting the radicals from the stem and filling the slots in the pattern. The shortcoming in this lemmatization approach is that it needs to be fed with the diacritized stem, the surrounding affixes and the POS tag to be able to return the correct lemma.

Farasa Segmenter/Lemmatizer [14] is a Support Vector Machine lemmatizer using a set of features to rank possible lemmas of a word. The features included in this supervised learning model are the likelihoods of stems, prefixes, suffixes, their combinations and presence in stems lexicon, stem patterns lexicon, gazetteer lexicon, function words lexicon, AraComLex lexicon and Buckwalter Lexicon. The drawback of this lemmatizer is the absence of diacritic marks in the output.

2.3 Hybrid Approach

Alkhalil lemmatizer [27] is a hybrid lemmatizer using both morphological analysis and statistical techniques to bring the lemma of the input text. Alkhalil lemmatizer relies in the first step on Alkhalil 2 to get the list of potential lemmas free of context and in the second step on the Hidden Markov Model (HMM) and the Viterbi algorithm to get the most appropriate lemma according to sentence context.

2.4 Summary

To sum up, in this literature review we presented attempts carrying out Arabic lemmatization. MADAMIRA, AlKhalil 2 and Alkhalil lemmatizer suffer from the first drawback which is the lemmatization through the morphological analysis, whereas Farasa Segmenter fails to provide diacritized output.

Our work differs from the previous ones. We propose a resource-based lemmatization approach, offering to the ANLP community a lexicon-based lemmatizer which provides diacritized output lemmas according to the input sentence context without involving any morphological analysis. To the best of our knowledge, this is the first attempt to address lemmatization using a lemmas lexicon.

3 Our Approach

The state of the art about the existing Arabic lemmatizers revealed some drawbacks when using the three aforesaid approaches (morphological, statistical and the hybrid). Our approach is different and takes its roots from the Arabic language itself.

Indeed, Arabic is a highly structured language respecting a templatic and concatenative morphology. The templatic feature refers to the morphology based on roots and patterns (templates) to identify the lemmas, while the concatenative one denotes that the lemma concatenates to affixes to produce the stem, and this latter concatenates to clitics to yield the word. Therefore, these two Arabic language features lead to a finite set of lemmas, stems and then words produced by applying a number of

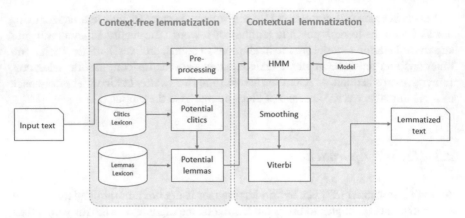

Fig. 1 SAFAR Lemmatizer architecture

well-defined rules. This Finiteness of Arabic lemmas/stems/words drove us to adopt a lexicon-driven approach.

Consequently, our approach consists in the adoption of the classical methodology with direct access to lexicons, because by storing the complete set of possible clitics and stems/lemmas, the lemmatizer will produce negligible errors. To this end, we use a clitics lexicon and a stem/lemma lexicon allowing us to get all acceptable lemmas (potential) for each word in the input text.

Figure 1 illustrates the architecture of our lemmatizer (named Safar lemmatizer). The user inputs a sentence to get the corresponding lemma of every word in the sentence. The lemmatization is processed in two steps. The first one is the context-free lemmatization, consisting in getting all acceptable lemmas (potential) using the clitics lexicon and a large stem/lemma lexicon. In the second step, we detect the most appropriate lemma for every word according to the sentence context using a supervised learning technique. Before detailing the two steps, we introduce the lexicon resource preparation leading to the clitics and the stem/lemma lexicons.

3.1 Lexicon Resource Preparation

The first lexicon is about clitics that has already been developed in a previous work [28]. Arabic clitics attach to the inflected base word (the stem) without any orthographic marks (like the possessive apostrophe in English). Thus, we define a clitic as a proclitic-enclitic couple, knowing that proclitics are clitics concatenated before the stem, and those concatenated after the stem are enclitics. For example, the word "وَبِكُتُبِهِمْ"(And with their books) is segmented as "وَبِكُتُبِهِمْ = وَبِ(and with) + كُتُب(books) + هِمْ(their)" denoting "وَبِ"as a proclitic and "هِمْ"as an enclitic.

The second lexicon is about stem/lemmas which must cover a large list of Arabic inflected word forms (stems) and their corresponding lemmas. The literature review related to the available lexicons offering the inflected-word/lemma pairs discloses the existence of some lexicons such as DIINAR [29], Arabic dictionary of inflected words,[2] ALIF [30], Arabic LDB [31] and Arabic verb resource [32]. The last two resources mainly deal with verbs, while DIINAR and Arabic dictionary of inflected words involve nominal and verbal lexical categories. However, these two resources are not freely available. Therefore, we used the data extracted from the Arabic Lexicon of Inflected Forms (ALIF) [30] that we enriched with missing lemmas.

Indeed, ALIF is a morpho-syntactic lexicon of inflected forms of the Arabic language in which each inflected form is associated with a morpho-syntactic information. In ALIF, a verb is the origin of the majority of the other categories, the inflection "التصريف"gives the conjugated forms of the verb and the derivation "الاشتقاق"gives the derived nouns. Every Arabic verb is formed from a triliteral or a quadriliteral root. According to the nature of the root letters, we distinguish two main classes: regular verbs (الصحيح)and irregular or weak verbs (المعتل). For example, the regular class includes sub classes such as pure or sane (السالم), doubled (مضعف)or hamzated (مهموز). Whereas the regular class includes sub classes such as assimilated (المثال)hollow (الأجوف)or doubly weak (اللفيف). Afterwards, the derived nouns are obtained through the derivation of all verbal categories. Accordingly, ALIF includes all Arabic verbs and derived nouns, but excludes Arabic particles and non-derived nouns (Primitive nouns). These omitted categories have been subject of an enrichment (as explained below) to cover all Arabic word categories.

ALIF is constructed from a data base of 7522 roots. The inflection and derivation of these roots gave rise to 7,099,562 stems representing 137,721 lemmas. Figure 2 is an extract of some ALIF entries. For instance, in, the stem "أبأتُ"(I did) admits the lemma "أبأ"(>aba>), the root "ءبء"('b'), the type "12" corresponding to "فعل ماضي مبني للمعلوم"(a perfect active verb), the POS "55" denoting " مفرد محايد متعد فعل ثلاثي مجرد"(a trilateral singular neutral transitive verb), the pattern "فَعَلْتُ"(faEalotu), the empty prefix and the suffix "تُ"(tu).

Stem	Lemma	Root	Type	POS	Pattern	Prefix	Suffix
أبَأتُ	أبَأ	ءبء	12	55	فَعَلْتُ	#	تُ
أبأنَا	أبَأ	ءبء	12	56	فَعَلْنَا	#	نَا
أبأنَا	أبَأ	ءبء	12	57	فَعَلْنَا	#	نَا
أبَأتَ	أبَأ	ءبء	12	58	فَعَلْتَ	#	تَ
أبَأتِ	أبَأ	ءبء	12	59	فَعَلْتِ	#	تِ
أبَأتُمَا	أبَأ	ءبء	12	60	فَعَلْتُمَا	#	تُمَا
أبَأتُمَا	أبَأ	ءبء	12	61	فَعَلْتُمَا	#	تُمَا
أبَأتُمْ	أبَأ	ءبء	12	62	فَعَلْتُمْ	#	تُمْ
أبَأتُنَّ	أبَأ	ءبء	12	63	فَعَلْتُنَّ	#	تُنَّ

Fig. 2 ALIF lexicon extract

---

[2]http://catalog.elra.info/en-us/repository/browse/ELRA-L0098/ Retrieved March 28, 2018.

Table 1 Arabic lemmas and stems statistics

	SAMA	Arabic lexicon (lsAnAlErb) "لسان العرب"	Arabic Lexicon (tAjAlErws) "تاج العروس"	ALIF	Stems/lemmas lexicon
Number of Lemmas	40,654	80,000	120,000	137,721	164,272
Number of Stems	79,318	–	–	7,099,562	7,133,106

From ALIF, we extracted the stem/lemma pairs which serve as a basis for the building of our stems/lemmas lexicon. As a matter of fact, the main need of this lexicon-based approach is the lexicon coverage. To this end, we enriched it with missing lemmas such as Arabic particles, broken plurals and omitted nouns. We added to the lexicon 720 Arabic particles taken from particles lexicon [28], 15,291 broken plurals manually compiled from Arabic dictionaries and 17,533 nouns from three nominal classes which are "نسبة"(Relative adjective), "مصدر صناعي"(Industrial verbal noun) and "اسم جامد"non-derivative nouns. This enrichment effort resulted in the enlargement of the lexicon with 26,551 additional lemmas to achieve the number of 164,272 lemmas representing 7,133,106 stems.

Compared to the number of Arabic lemmas and stems found in the literature, and to the best of our knowledge (Table 1), our lexicon (with 164,272 lemmas and more than 7 million stems) can be considered as the most comprehensive lemmas lexicon. This lexicon represents the key resource in the context-free lemmatization that extracts the potential lemmas to be used in the contextual lemmatization step.

3.2 Context-Free Lemmatization

In the pre-processing step we normalize the input text by removing non Arabic and numerical letters. We don't remove stop words in the pre-processing step because our stems/lemmas lexicon contains also stop words (720 Arabic particles "Sec. 3.1"). So, stop words are treated as any other word. By doing so, we maintain an undamaged context, while the stop words removal will damage the overall context, what will affect the lemmatizer performances in the contextual lemmatization phase.

After that, the context-free lemmatization step is divided into two sub-processes.

The first sub-process is the clitics extraction. In this step, SAFAR lemmatizer detects all possible combinations of clitics through querying the "clitics lexicon" that returns a list of proclitic-enclitic couples. Then, the obtained list of clitics is filtered to generate a list of probable stems. As the example illustrated in Fig. 3 shows, the input word "سيكتبهم"(syktbhm)[3] provides the following four potential

---

[3]In this stage of processing, we could not provide a translation for the word because the word is ambiguous.

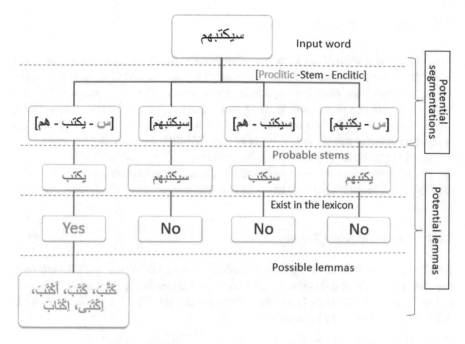

Fig. 3 SAFAR context-free lemmatization

segmentations (م - يكتب - هم} ، } - سيكتبهم - { ، } - سيكتب - هم{ ، } - يكتبهم{ سُ}).
From these segmentations, the potential clitic {هم-سُ}gives the probable stem
"يكتب"(yktb) (سيكتبهم = س+يكتب+هم)and so on.

After this step comes the role of the lexicon resource "stem/lemma lexicon" which
acts as a validator of the obtained probable stems list to give the possible lemmas. If
the probable stem exists in the "stem/lemma lexicon", we consider the corresponding
lemma to be a possible one for the input word. Otherwise, the non-existence of
the probable stem in our lexicon resource means that the segmentation giving this
probable stem is invalid.

For example, by checking the existence of the probable stem "يكتب"(yktb) in
the lexicon, we obtain five possible lemmas "كَتَّبَ، كَتَبَ، أَكْتَبَ، اِكْتَبَى، اِكْتَابَ"(make
write, write, dictate, bend, learn to write). While using the potential segmentation
{ - يكتبهم - سُ}, checking the existence of the probable stem "يكتبهم"(yktbhm) gives
an empty result indicating that it is not a correct stem.

In contrast, in order to cover some categories of words that we cannot inventory
(such as named entities), the lemma of a misrecognized input words (inputs that
cannot be segmented or out of lexicon) is itself.

3.3   Contextual Lemmatization

After the identification of the list of potential lemmas for the input text, the contextual
lemmatization step consists of removing the ambiguity related to the correct lemma
among the potential ones (as explained previously in the introduction). For this
purpose, we apply a supervised learning technique to identify, for every input word,
the most appropriate lemma according to the sentence context. This statistical step
is conducted using the Hidden Markov Model (HMM), a smoothing technique and
the Viterbi algorithm. This choice can be explained by the fact that HMM technique
has already proved its worth in this field [27] and the evaluation comparing it with
the Language Model (LM) [33] (discussed in the next section) support this choice.

3.3.1   Hidden Markov Model

An HMM [34] represents a sequence of two random variables; the first is hidden and
the second is observed. HMM, as defined in books dealing with statistics, is used
to predict the hidden states from the observed ones. An HMM $\lambda = (S, A, B, \pi)$ is
defined by the following parameters:

S: the set of possible states $S = \{s_1, s_2, \ldots, s_f\}$.
A: Matrix of transition between states, where the element $a(i, j)$ is the transition
probability of the process from a state i (at the time t $-$ 1) to a state j (at the time t).
It is defined by: $a(i, j) = P(X_t = s_j | X_{t-1} = s_i)$.
B: Matrix of emission defined by the elements: $b(i, j) = b_i(j) = P(O_t = s_j | X_t = s_i)$, where: $X = X_1, X_2, \ldots, X_e$ are the hidden states and $O = O_1, O_2, \ldots, O_e$ are
the observed states.
π: The initial states vector of the process $\pi = \{\pi_i\}_{i=1,\ldots,K}$; where π_i represents the
probability that the state at the time $t = 1$ is i, defined by: $\pi(i) = P(X_1 = s_i)$.

By applying HMM in our context, to find for the sentence $Ph = (w_1, w_2, \ldots, w_n)$
the most probable sequence of lemmas $(l_1^*, l_2^*, \ldots, l_n^*)$, we estimate the model parameters using a training corpus C (composed by N words and M sentences) and the
maximum likelihood estimation method [34]. In other terms, in our HMM model,
the observed states are the words in the input sentence and the hidden ones are the
possible lemmas obtained in the context-free lemmatization step. The HMM model
$\lambda = (S, A, B, \pi)$ admits the following parameters:

$S = \{l_1, l_2, \ldots, l_m\}$ is the set of lemmas in the Arabic language
$a(i, j) = \frac{n_{ij}}{n_i}, 1 \leq i \leq N, 1 \leq j \leq N$, the probability for a lemma l_i to be followed
by the lemma l_j
$b_i(t) = \frac{m_{it}}{n_i}, 1 \leq i \leq N, 1 \leq t \leq n$, the probability for the word w_t to give the
lemma l_i
$\pi_i = \frac{n_{io}}{M}$, the probability for the sentence to start with lemma l_i
where:
n_{io}: occurrences number in C of sentences starting with l_i

n_{ij}: occurrences number in C of the lemma l_i followed by the lemma l_j
m_{it}: occurrences number in C of the word w_t associated with the lemma l_i
n_i: occurrences number in C of the lemma l_i.

3.3.2 Smoothing

Since there is no training corpus containing all the transitions and emission of the words in the Arabic language, the elements of the matrices A and B can be estimated for several words by the value zero. This will negatively affect the optimal path search by the Viterbi algorithm. To adjust this irregularity, smoothing techniques are then used. These techniques consist in assigning a non-zero probability to the elements of the matrices A and B during the test phase of the HMM model. For this, we use the Absolute Discounting method [35] with the same logic as in [27] except for the parameter Z_i.

Then, the HMM parameters π_i, a_{ij} and $b_i(t)$ are estimated by:

$$\pi_i = \frac{\max(m_i - 0.5, 0)}{M} + \left(\frac{0.5}{M} \times \frac{n_j}{N} \times N_{i+} \right)$$

$$a_{ij} = \frac{\max(n_{ij} - 0.5, 0)}{n_i} + \left(\frac{0.5}{n_i} \times \frac{n_j}{N} \times N_{li+} \right)$$

$$b_i(t) = \begin{cases} \frac{m_{it} - 0.5}{n_i}, & \text{if } m_{it} \neq 0 \\ \frac{N \times 0.5}{n_i \times Z_i}, & \text{elsewhere} \end{cases}$$

with

N_{i+}: occurrences number in C of words for which the lemma appears in the beginning of a sentence
N_{li+}: occurrences number in C of words for which the lemma appears after l_i
Z_i: occurrences number in C of words not annotated with l_i but obtained in the context-free phase.

3.3.3 Viterbi

The Viterbi algorithm [36] provides an optimal maximum likelihood solution for the estimation of a sequence of states of a HMM. In other words, given an HMM λ and a set of observations (a sentence Ph $= (w_1, w_2, \ldots, w_n)$), we use the Viterbi algorithm to determine the most probable states sequence (sequence of lemmas $(l_1^*, l_2^*, \ldots, l_n^*)$) engendering Ph. To do so, we compute the joint probability of the observation sequence with the best state sequence as bellow:

$$\text{Initiation:} \quad \delta_1(j) = a_{0j}\, b_j(w_1)$$

$$\psi_1(j) = 0$$

$$\text{Recursion:} \quad \delta_t(j) = \max_{i=1,..,N} \delta_{t-1}(i)\, a_{ij}\, b_j(w_t)$$

$$\psi_t(j) = \operatorname{argmax}_{i=1,..,N} \delta_{t-1}(i)\, a_{ij}\, b_j(w_t)$$

$$\text{Termination:} \quad P^* = \delta_t(j) = \max_{i=1,..,N} \delta_n(i)\, a_{in}$$

$$l_n^* = \psi_n(l_n) = \operatorname{argmax}_{i=1,..,N} \delta_n(i)\, a_{in}$$

Fig. 4 Viterbi algorithm steps

$$\delta_t(j) = \max_{s_0, s_1, \ldots, s_{t-1}} P(s_0, s_1, \ldots, s_{t-1}, w_1, w_2, \ldots, w_t, s_t = j | \lambda)$$

It can be written:

$$\delta_t(j) = \max_{i=1,\ldots,N} \delta_{t-1}(i) a_{ij} b_j(w_t)$$

To identify the most probable states sequence for our HMM, we implemented the algorithm steps (initiation, recursion and termination) illustrated in Fig. 4. The "initiation" consists to give the best probability to the first word of the sentence. The "recursion" determines recursively the best path (the maximum probability of a hidden states sequence) leading to every word in the sentence. Finally, the "termination" involves the backtrack of the best path.

As illustrated in Fig. 5, the input words represent the observed states in our HMM, and the list of potential lemmas provided by the context-free lemmatization step for every input word represents the hidden states. The contextual lemmatization consists firstly, in the estimation of the HMM parameters by applying the smoothing technique, namely the initial states vector π, the emission matrix B and the transition matrix A. Secondly, we apply the Viterbi algorithm to identify the most probable sequence of lemmas representing the best lemmas according to the input sentence.

To check the effectiveness of our lemmatization approach we evaluate it and compare our lemmatizer with the existing ones.

4 Evaluation and Comparison

To demonstrate the efficiency of SAFAR lemmatizer, we evaluate the context-free and the contextual lemmatization steps, then we compare our lemmatizer with those of the literature. After that, we experiment the lemmatization effects on improving the efficiency of a document classification task.

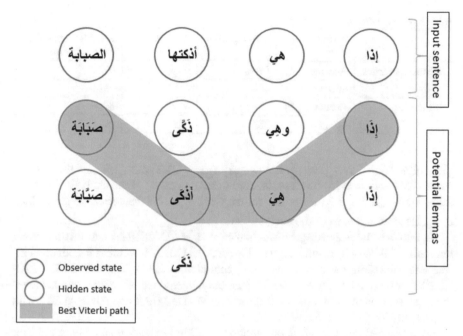

Fig. 5 Contextual lemmatization example

4.1 Context-Free Evaluation

This evaluation consists in assessing the results of the context-free lemmatization with the outcome of the two known morphological analyzers MADAMIRA and ALKHALIL 2 since they offer all possible lemmas of a given word. To achieve this, we use two annotated corpora, namely Al-Mus'haf corpus [37] and Nemlar [38]. These corpora are useful since they contain the lemma tag in their annotation and are freely available. They are formatted in plain text where the verses (for Al-Mus'haf corpus) and sentences (for Nemlar) are written in separate lines. Every line is a set of pairs word/lemma separated by a white space.

The evaluation consists in checking for the two corpora, if the analysis result contains the correct lemma or not, using the accuracy (%) and the speed (words/second) metrics:

$$\text{Accuracy} = \frac{\text{Number of words for which the correct lemma exists in the analysis results}}{\text{Number of words in the corpus}}$$

$$\text{Speed} = \frac{\text{Number of words in the corpus}}{\text{Processing time}}$$

For example, if the analysis of the sentence "المبدأ الأساسي في العمل" gives the following output list

Table 2 Context-free lemmatization evaluation

	Metric	MADAMIRA	AlKHALIL2	SAFAR context-free lemmatization
Al-Mus'haf corpus	Accuracy (%)	67.94	91.41	98.36
	Speed (w/s)	358	671	2816
NEMLAR	Accuracy (%)	78.63	61.17	86.03
	Speed (w/s)	382	693	2981

[(مِبْدَأً، مُبْدَأً، مُبْدَأً، مَبْدَأً); (أَسَاسٌ، أَسَاسِيٌّ، الأَسَاسِيُّ); (فِي، وَفَى،); (عَمَلٌ، عَمَّلَ عَمِلَ،)], the accuracy is 75% (3/4 = 0.75) because the correct lemma exists in the analysis list of the three words "العمل، في، المبدأ", while for the word "الأساسي" the correct lemma which is "أَسَاسِيٌّ" does not exist in the analysis list.

As shown in Table 2, using Al-Mus'haf corpus, MADAMIRA includes the correct lemma in 67.94% of its results and 91.41% for AlKHALIL 2, whereas for our context-free lemmatization, the correct lemma is included in the output of 98.36% of Al-Mus'haf corpus. Also, for Nemlar Corpus, our context-free lemmatization achieves the greatest accuracy with 86.03% followed by MADAMIRA and AlKHALIL 2 with respectively 78.63% and 61.17%.

From a speed point of view, our context-free lemmatization step is around four times faster than AlKHALIL 2 and eight times faster than MADAMIRA.

The context-free step evaluation demonstrates that the adopted approach provides higher results compared to MADAMIRA and AlKHALIL 2 for both accuracy and processing speed metrics. The second observation is that the maximal accuracy obtained using NEMLAR corpus is 86.03%. As a result, we ask questions about the remaining 13.97%. A quick survey of these entries demonstrates that the corpus contains some errors as shown in the Table 3:

And as consequence, we limit our future evaluation and comparison to Al-Mus'haf corpus.

To evaluate the context detection step, we compare the supervised learning technique adopted in this phase (HMM) with the Language Model (LM) technique.

4.2 Context Detection Evaluation

To build our classifier models, we use 90% of Al-Mus'haf corpus for training and the remaining 10% for test purposes. Thus, we implement the LM technique by employing the Kyoto Language Modeling Toolkit[4] for both mono and bi-grams, and we compare the accuracy and processing speed obtained by the three learning techniques. Table 4 shows that the LM Mono-gram is the best one in terms of processing speed with 1996 words processed per second, followed by LM Bi-gram and HMM.

---

[4]http://www.phontron.com/kylm/ Retrieved March 28, 2018.

Table 3 Examples of errors in NEMLAR corpus

Word	Lemma	Correct word	Correct lemma	Error
كَلَمَةٌ	-	كَلِمَةٌ	كَلِمَة	Incorrect word
لِأهبَ	-	لِأهَبَ	وَهَبَ	Incorrect word
مُزيجٌ	-	مَزيجٌ	مَزيج	Incorrect word
اَلفِقهِ	-	اَلفِقهِ	فِقه	Incorrect word
لِلأجُورِ	أجُوَر	لِلأجُورِ	أجر	Incorrect word
أجمَعِينَ	جَمِيع	أجمَعِينَ	أجمَع	Incorrect lemma
هُمَا	هُوَ	هُمَا	هُمَا	Incorrect lemma
وَتِلكَ	ذَا	وَتِلكَ	تِلكَ	Incorrect lemma
إمَّا	إن	إمَّا	إمَّا	Incorrect lemma
أبَالِسَةٌ	أبَالِسَة	أبَالِسَةٌ	إبلِيس	Incorrect lemma
أجَاوِيذَ	أجَاوِيد	أجَاوِيذَ	جَواد	Incorrect lemma
يَأتُونَهُ	-	يَأتُونَهُ	أتَف	Unknown lemma
اَلطَعُونِ	-	اَلطَعُونِ	طَعُن	Unknown lemma
وَلِلجُنَةِ	-	وَلِلجُنَةِ	لَجُنَة	Unknown lemma
أمَامُهُ	-	أمَامُهُ	أمَام	Unknown lemma
اَلعَقَارَاتِ	-	اَلعَقَارَاتِ	عَقَار	Unknown lemma

Table 4 Context detection evaluation

	LM Mono-gram	LM Bi-gram	HMM
Accuracy (%)	58.92	67.79	91.31
Speed (w/s)	1996	1597	887

However, in terms of accuracy HMM advances the mono and the bi-gram models with 91.31% of accuracy. Subsequently, this comparison leads to the adoption of the HMM as a context detection technique. Even if it is the slowest one, the accuracy gap between HMM and the two LMs cannot be neglected (32.39% with the Mono-gram and 23.52% with the Bi-gram).

4.3 Lemmatizers Comparison

We compare in this stage the performance of SAFAR lemmatizer with the existing lemmatization tools MADAMIRA, AlKHALIL lemmatizer and FARASA Segmenter. The training set used for SAFAR lemmatizer is 90% of Al-Mus'haf corpus and the corpus used to test the four lemmatization tools is 10% of Al-Mus'haf corpus. As shown in Table 5, SAFAR lemmatizer achieves the best accuracy with 91.31%, while FARASA Segmenter is the fastest.

FARASA Segmenter provides an undiacritized output. That means the output remains ambiguous; for instance, the lemma "كتب"(ktb) can be "كَتَبَ، كَتَّبَ، كُتِب"(to write, make write, written). Accordingly, the 83.56% of accuracy obtained by

Table 5 lemmatization tools comparison using Al-Mus'haf corpus

	MADAMIRA	AlKHALIL Lemmatizer	FARASA Segmenter	SAFAR Lemmatizer
Accuracy (%)	57.94	83.14	83.56	91.31
Speed (w/s)	319	666	998	887

Table 6 lemmatization tools comparison using newspaper articles

	Words number	Accuracy (%)/Speed (w/s)		
		MADAMIRA	AlKHALIL Lemmatizer	SAFAR Lemmatizer
Politics	653	50.00/328	67.83/674	85.69/849
Art	736	46.85/334	58.16/681	79.29/857
Sport	315	22.61/306	30.19/637	83.63/836
Economy	858	50.82/352	60.37/692	77.29/898
AVG	641	42.57/330	54.14/671	81.48/860

FARASA cannot be considered because it is a biased value and speed alone is not a reliable indicator of the overall efficiency.

In addition, since Quranic text is considered by some researchers as a classical Arabic, we decided to make the comparison using another corpus by gathering articles in four domains from an electronic newspaper site.[5] We manually annotated the new corpus with the lemma tag. Table 6 describes the collected articles and the obtained accuracy using the three tools. From the obtained results, we notice that for all domains, the lemmatizer having the greatest accuracy and speed is SAFAR followed by AlKHALIL and MADAMIRA.

4.4   Experiment

As already mentioned in the introduction, lemmatization reduces indexing data dimension more than stemming. To prove the truthfulness of this statement, in this experiment, we observe the lemmatization effects on improving the efficiency of a Text Classification (TC) task. Its consists to classify documents into a set of prede-fined categories using words, roots (using Khoja root-based stemmer [39]), stems (using Saad stemmer [40]) and lemmas (using SAFAR lemmatizer). Khoja root-based and Saad stemmers choice is done after an evaluation of stemmers performances [41].

The experiment was performed on a database of 144 Arabic documents collected from an electronic news paper (see footnote 5) and classified in four equally dis-tributed categories (Art, Economics, Politics and Sports). This dataset is divided into

---

[5] www.hespress.com.

two sub sets, 100 documents are used for Training (69.5%) and 44 documents for testing (30.5%). The experiment consists of classifying the dataset documents using the Support Vector Machine (SVM) machine learning algorithm. The Feature vector dimension metric is used to evaluate the indexing data dimension efficiency, while F1 measure, correctly classified instances and incorrectly classified instances are used to assess the TC accuracy.

Accordingly, the analysis of the average results obtained in Table 7 demonstrates that, firstly, lemmatization improves the TC performances since the average F1 measure and the average Correctly Classified documents for the lemmatized classifier outperforms the average of the non-lemmatized ones. Secondly, the classification using Safar lemmas outperforms the one made using AlKhalil (obviously "according

Table 7 SVM text classification results

Classification with	Category	Feature vector dimension	F1 (%)	Correctly classified (%)	Incorrectly classified (%)
words	Art	13,613	50.00	36.36	63.64
	Economy	8384	83.30	90.91	9.09
	Politics	4826	100.00	100.00	0.00
	Sport	3820	76.90	90.91	9.09
	Average	**7660.75**	**77.60**	**79.55**	**20.45**
Khoja Roots	Art	2888	53.3	36.36	63.64
	Economy	1876	95.7	100.00	0.00
	Politics	1635	91.7	100.00	0.00
	Sport	1286	76.9	90.90	9.09
	Average	**1537**	**79.4**	**81.81**	**18.19**
Saad Stems	Art	7928	62.5	45.45	54.54
	Economy	4523	90.9	90.90	9.09
	Politics	3565	95.7	100.00	0.00
	Sport	2619	81.5	100.00	0.00
	Average	**4658.75**	**82.6**	**84.10**	**15.90**
Alkhalil lemmas	Art	7553	84.20	72.73	27.27
	Economy	3861	90.90	90.91	9.09
	Politics	3027	95.70	100.00	0.00
	Sport	2269	91.70	100.00	0.00
	Average	**4177.5**	**90.60**	**90.91**	**9.09**
SAFAR lemmas	Art	6287	90.00	81.82	18.18
	Economy	3845	95.70	100.00	0.00
	Politics	3032	100.00	100.00	0.00
	Sport	2272	95.70	100.00	0.00
	Average	**3859**	**95.30**	**95.45**	**4.55**

to lemmatization tools comparison" with that made using MADAMIRA). Thirdly, lemmatization improves the TC efficiency by reducing the Feature vector dimension. By comparing feature vector dimension, we mention that roots-based classification (1537) is more efficient than lemmas-based one (3859). However, from a performance point of view (F1 and Correctly Classified documents), the accuracy of lemmas-based classification outperforms roots-based one.

Finally, from all evaluations and comparisons above, we confirm that SAFAR lemmatizer outperforms all other tools in terms of accuracy, processing speed and data dimension efficiency.

5   Conclusion

In this paper, we proposed a new lemmatizer offering diacritized lemmas (not stems) of the input text without relying on the morphological analysis. SAFAR lemmatizer provides the most probable lemmas of an input sentence, first by getting all acceptable lemmas for every input word using the stem/lemma lexicon, second by getting the most appropriate lemma to sentence context based on the hidden Markov model.

The context-free evaluation has led to more than 98% of accuracy and the contextual steps has showed that the HMM is the best context detection technique. The conducted comparison provides SAFAR Lemmatizer with a leading position in contrast with the state of the art lemmatizers.

In the future, we intent to improve our lemmatizer along two main lines:

- Stem/lemma lexicon enrichment with missing lemmas such as named entities
- Context detection enhancement to reduce the gap (7.05%) between the context-free and contextual lemmatization.

Finally, we note that SAFAR lemmatizer will be available soon and a demo version is accessible via the Safar web plateform.[6]

References

1. The World Bank, *World Development Indicators* (The World Bank, Washington, DC, 2018)
2. J. Owens, *The Oxford Handbook of Arabic Linguistics* (Oxford University Press, 2013), p. 2
3. R. Zhang, E. Sumita, Boosting statistical machine translation by lemmatization and linear interpolation, in *Proceedings of the 45th Annual Meeting of the ACL on Interactive Poster and Demonstration Sessions. Association for Computational Linguistics* (2007)
4. K. Tuomo, et al., Stemming and lemmatization in the clustering of finnish text documents, in *Proceedings of the Thirteenth ACM International Conference on Information and Knowledge Management* (ACM, 2004)
5. E.-S. Tarek, F. El-Ghannam, *A Lemma Based Evaluator for Semitic Language Text Summarization Systems*. arXiv preprint arXiv:1403.5596 (2014)

---

[6]http://arabic.emi.ac.ma:8080/SafarWeb_V2/faces/safar/morphology/lemmatizer.xhtml.

6. E.-S. Tarek, A. Al-Sammak, Arabic Keyphrase Extraction Using Linguistic Knowledge and Machine Learning Techniques. arXiv preprint arXiv:1203.4605 (2012)
7. G. De Pauw, G.-M. De Schryver, Improving the computational morphological analysis of a Swahili corpus for lexicographic purposes. Lexikos **18**(1) (2008)
8. M.N. Al-Kabi et al., A novel root based Arabic stemmer. J. King Saud Univ.-Comput. Inf. Sci. **27**(2), 94–103 (2015)
9. M. Al-Kabi, R. Al-Mustafa, Arabic root based stemmer, in *Proceedings of the International Arab Conference on Information Technology*, Jordan (2006)
10. L.S. Larkey, L. Ballesteros, M.E. Connell, *Light Stemming for Arabic Information Retrieval. Arabic Computational Morphology* (Springer, Dordrecht, 2007), pp. 221–243
11. P. Arfath et al., Madamira: a fast, comprehensive tool for morphological analysis and disambiguation of arabic. *LREC* **14** (2014)
12. D. Taji, S. Khalifa, O. Obeid, F. Eryani, N. Habash, An Arabic morphological analyzer and generator with copious features, in *Workshop on Computational Research in Phonetics, Phonology, and Morphology*. The Conference on Empirical Methods in Natural Language Processing (EMNLP 2018), Brussels, Belgium
13. B. Mohamed et al., AlKhalil Morpho Sys 2: a robust Arabic morpho-syntactic analyzer. J. King Saud Univ.-Comput. Inf. Sci. **29**(2), 141–146 (2017)
14. A. Ahmed, et al., Farasa: a fast and furious segmenter for arabic, in *Proceedings of the 2016 Conference of the North American Chapter of the Association for Computational Linguistics: Demonstrations* (2016)
15. E. Al-Shammari, J. Lin, A novel Arabic lemmatization algorithm, in *Proceedings of the Second Workshop on Analytics for Noisy Unstructured Text Data* (ACM, 2008)
16. B. Karien, R. Pretorius, G.B. van Huyssteen, Automatic lemmatization in Setswana: towards a prototype. S. Afr. J. Afr. Lang. **25**(1), 37–47 (2005)
17. K.R. Beesley, Finite-state morphological analysis and generation of Arabic at xerox research: status and plans in 2001, in *ACL Workshop on Arabic Language Processing: Status and Perspective*, vol 1 (2001)
18. B. Tim, Issues in Arabic orthography and morphology analysis, in *Proceedings of the Workshop on Computational Approaches to Arabic Script-Based Languages*. Association for Computational Linguistics (2004)
19. S. Otakar, Elixirfm: implementation of functional arabic morphology, in *Proceedings of the 2007 Workshop on Computational Approaches to Semitic Languages: Common Issues and Resources*. Association for Computational Linguistics (2007)
20. M. Altabba, A. Al-Zaraee, M.A. Shukairy, An Arabic morphological analyzer and part-of-speech tagger. Actes de JADT (2010)
21. S. Majdi, E. Atwell, M.A.M. Abushariah, SALMA: standard Arabic language morphological analysis, in *2013 1st International Conference on Communications, Signal Processing, and their Applications (ICCSPA)*. IEEE (2013)
22. B. Abderrahim, et al., Alkhalil morpho sys1: A morphosyntactic analysis system for arabic texts, in *International Arab Conference on Information Technology*. Benghazi Libya (2010)
23. H. Nizar, O. Rambow, R. Roth, MADA + TOKAN: a toolkit for Arabic tokenization, diacritization, morphological disambiguation, POS tagging, stemming and lemmatization, in *Proceedings of the 2nd International Conference on Arabic Language Resources and Tools (MEDAR)*, Cairo, Egypt, vol 41 (2009)
24. D. Mona, Second generation AMIRA tools for Arabic processing: fast and robust tokenization, POS tagging, and base phrase chunking, in *2nd International Conference on Arabic Language Resources and Tools*, vol 110 (2009)
25. D. Graff, M. Maamouri, B. Bouziri, S. Krouna, S. Kulick, T. Buckwalter, *Standard Arabic Morphological Analyzer (SAMA) Version 3.1. Linguistic Data Consortium LDC2009E73* (2009)
26. A. Mohammed, A. Zirikly, M. Diab, The power of language music: Arabic lemmatization through patterns, in *Proceedings of the 5th Workshop on Cognitive Aspects of the Lexicon (CogALex-V)* (2016)

27. B. Mohamed, M. Azzeddine, Approche hybride pour le développement d'un lemmatiseur pour la langue arabe. 13éme Colloque Africain sur la Recherche en Informatique et Mathématiques Appliquées (2016)
28. D. Namly, Y. Regragui, K. Bouzoubaa, Interoperable Arabic language resources building and exploitation in SAFAR platform, in *2016 IEEE/ACS 13th International Conference of Computer Systems and Applications (AICCSA)*. IEEE (2016)
29. J. Dichy, M. Hassoun, The DIINAR. 1 Arabic Lexical Resource, an outline of contents and methodology. The ELRA Newsl. **10**(2) (2005)
30. A. El Jihad, D. Namly, K. Bouzoubaa, The development of a standard Morpho-Syntactic Lexicon for Arabic NLP, in *Proceedings of the International Conference on Learning and Optimization Algorithms: Theory and Applications* (ACM, 2018)
31. A. Khemakhem, Arabic LDB: A Standardized Lexical Basis for the Arabic Language (2006)
32. A. Neme, A fully inflected Arabic verb resource constructed from a lexicon of lemmas by using finite-state transducers. Revue RIST **20**(2), 7–19 (2013)
33. F. Song, B. Croft, A general language model for information retrieval, in *Proceedings of the Eighth International Conference on Information and Knowledge Management* (ACM, 1999)
34. O. Ibe, *Markov Processes for Stochastic Modeling*. Elsevier insights, Elsevier Science, 2nd edn (2013)
35. H. Ney, U. Essen, On smoothing techniques for bigram-based natural language modelling, in *[Proceedings] ICASSP 91: 1991 International Conference on Acoustics, Speech, and Signal Processing, Toronto, Ontario, Canada*, vol 2 (1991), pp. 825–828
36. D. Forney, The Viterbi algorithm. Proc. IEEE **61**(3), 268–278 (1973)
37. I. Zeroual, A. Lakhouaja, A new Quranic Corpus rich in morphosyntactical information. Int. J. Speech Technol. (IJST) (2016)
38. M. Boudchiche, A. Mazroui, Enrichment of the Nemlar corpus by the lemma tag, in *Workshop Language Resources of Arabic NLP: Construction, Standardization, Management and Exploitation. Rabat, Morocco*. November 26 (2015)
39. K. Shereen, R. Garside, Stemming arabic text. Lancaster, UK, Computing Department, Lancaster University (1999)
40. M. Saad, W. Ashour, *Arabic Morphological Tools for Text Mining* 18 (2010)
41. Y. Jaafar, D. Namly, K. Bouzoubaa, A. Yousfi, *Enhancing Arabic Stemming Process Using Resources and Benchmarking Tool*, King Saud University - Computer and Information Sciences (JKSU-CIS) 12/ 2016

The Role of Transliteration in the Process of Arabizi Translation/Sentiment Analysis

Imane Guellil, Faical Azouaou, Fodil Benali, Ala Eddine Hachani
and Marcelo Mendoza

Abstract Arabizi is a form of written Arabic which relies on Latin letters, numerals and punctuation rather than Arabic letters. In literature most of the works are concentrated in the study of Arabic neglecting the study of Arabizi. To conduct automatic translation and sentiment analysis, some approaches tend to handle it like any other language while others use a transliteration phase which converts Arabizi into Arabic script. In this context, the main purpose of this study is to determine the utility of Arabizi transliteration in improving automatic translation and sentiment analysis results. We introduce a rule-based automatic transliteration system. Then we apply this system to transliterate a collection of messages before proceeding to machine translation and sentiment analysis tasks. To evaluate the importance of transliteration on these tasks, we also present the construction of a set of lexical resources, such as: a parallel corpus between Arabizi and Modern Standard Arabic (MSA) constructed manually, a sentiment lexicon built automatically and revised manually, and an annotated sentiment corpus constructed automatically based on the sentiment lexicon. We also apply a set of algorithms and models dedicated to machine translation and sentiment analysis, including a number of shallow and deep classifiers as well as different embedding-based models for feature extraction. The experimental

I. Guellil (✉) · F. Azouaou
Laboratoire des Méthodes de Conception des Systèmes, Ecole nationale Supérieure
d'Informatique, BP 68M, 16309 Oued-Smar, Algiers, Algeria
e-mail: i_guellil@esi.dz

F. Azouaou
e-mail: f_azouaou@esi.dz

F. Benali
L'université de Versailles Saint Quentin, Versailles, France
e-mail: df_benali@esi.dz

A. E. Hachani
Sarl Services Web et Promotion (SWP), 12 Chemin Petit Hydra, El Biar, Algiers, Algeria
e-mail: da_hachani@esi.dz

M. Mendoza
Universidad Técnica Federico Santa María, Av. Vicuña Mackenna, 3939 Santiago, Chile
e-mail: marcelo.mendoza@usm.cl

© Springer Nature Switzerland AG 2020
M. Abd Elaziz et al. (eds.), *Recent Advances in NLP: The Case of Arabic
Language*, Studies in Computational Intelligence 874,
https://doi.org/10.1007/978-3-030-34614-0_6

101

results show a consistent improvement after applying transliteration achieving performance results up to 13.06 for automatic translation using the BLEU score and up to 92% for sentiment classification using the F1-score. This study allows to affirm that transliteration is a key factor in Arabizi handling.

Keywords Arabizi translation · Sentiment analysis · Machine translation

1 Introduction

Arabic is recognised as the 4th most used language of the Internet. Arabic has three main varieties: (1) classical Arabic (CA), (2) Modern Standard Arabic (MSA), (3) Arabic Dialect (AD). MSA and AD could be written either in Arabic or in Roman script (Arabizi), which corresponds to Arabic written with Latin letters, numerals and punctuation. Over the last decade, the Arabic and its dialects have begun to attract the attention of researchers in the area of natural language processing (NLP). Much work has been done to address different aspects of how this language and its dialects should be processed. Among these efforts stand out morphological analysis, lexical resource construction, and automatic translation, among other topics [1]. However, Arabic speakers in social media, short message systems (SMS), online chat applications and discussion forums often tend to use a non-standard romanization dubbed "Arabizi" [2, 3]. For example, the sentence "I'm happy today" is most commonly written in Arabizi ("rani fer7ana lyoum") than in Arabic (راني فرحانة ليوم). Arabizi is an Arabic text written using Latin characters, numerals and punctuation symbols [2].

A major challenge in the processing of Arabizi is that some words can be written in many different ways. For example, the word ان شَاء اللّه that means "if the god willing" can be written in Arabizi in 69 different manners [4]. Thus, the processing of texts written in Arabizi is challenging and difficult. We affirm that any system designed to analyse Arabic texts, especially in the case of text retrieved from social networks, it requires to include the processing of Arabizi messages. However, most of the work reviewed in the literature only focuses on Arabic scripting [5–13]. On the other hand, only a few works have been proposed to process Arabizi scripts [14–19].

The main idea of the works that deals with Arabizi is to introduce a transliteration system that is capable of converting Arabizi into Arabic. However, the transliteration process is challenging and difficult. For this reason, most of the works devoted to automatic translation or sentiment analysis are only focused on Arabic texts neglecting the processing of Arabizi [20–28]. Only a minor proportion of the works have been conducted in Arabizi automatic translation and sentiment analysis [29–35]. Some of these works first proceed to a transliteration phase [29–33] while the others handle Arabizi constructing annotated Arabizi corpus, neglecting the use of a transliteration system [34, 35].

The main purpose of this paper is to show if it is crucial to conduct an Arabizi transliteration phase before automatic translation, sentiment analysis. To answer this question, we propose a transliteration system based on a set of rules, that makes use of a statistical language model. Then we apply this system to a set of messages retrieved from social media. Finally, we conduct over the set of messages two tasks: machine translation and sentiment analysis. To validate our proposal, we evaluate the results in these two tasks with and without the use of our Arabizi transliteration system.

The main contributions of this work are the following:

- A rule-based transliteration system for Arabizi to Arabic is introduced. The system does not require annotated data.
- We introduce the first corpus of parallel sentences between Arabizi (Algerian dialect) and MSA. This corpus comprises 2924 messages retrieved from social media (Facebook and Twitter) and manually translated into MSA.
- We release a manually reviewed version of a sentiment lexicon automatically constructed by Guellil et al. [36]. This new version of the lexicon contains 1745 annotated words from which 968 are negatives, 771 are positives and 6 are neutral.
- An annotated sentiment corpus is introduced. This corpus was automatically built using the lexicon discussed above.
- We present an analytically study that evaluates the impact of the use of a transliteration system in Arabizi translation and sentiment analysis. Our results show that the use of transliteration is crucial for either translation and sentiment analysis.

2 Arabizi Transliteration Issues

The main problem of Arabizi language processing is related to the fact that one word can be written in many different manners. Cotterell et al. [4] argue that the word ان شَاء اللّه (meaning "if the god willing") can be written in 69 different manners such as: *InShalah, nshalah, nchalah, nshallah, nchallah, among others*. To face this challenge, almost all the research in the literature propose a transliteration process to transform from Arabizi to Arabic [15–17, 19, 29–33, 37]. Transliteration is just a transformation process from a written text in a given script to another. However, the transliteration from Arabizi to Arabic is not a simple task of replacing an Arabizi letter with one equivalent in Arabic. This task is a much more challenging task. To illustrate the real problem behind the Arabizi transliteration process, we show in Table 1 a list of messages in Arabizi along with their possible transliterations. The last column in Table 1 shows the correct transliteration of each sentence. Along this table it can be seen that some Arabizi letters (for example: a, e or t) can be equivalent to five or more Arabic letters. Sometimes, the transliteration process must consider a group of letters and not a letter individually (for example: sh and ch are equivalent to the Arabic letter ش).

After carefully analysing Table 1 we conclude that the most important issues related to Arabizi transliteration are:

Table 1 Transliteration examples from Arabizi to Arabic

Arabizi message	English translation	Possible transliteration	Correct transliteration
oumba3d nrouh	I will go after	ومبعد نروه أومبعد نره أومبعد نروح مبعد نرح	أومبعد نروح
ch7al chaba	How beatiful is she	كهحَال كهبَا شحَال شَابة شهل شبَا كهّال شَابَا	شحَال شَابة
ktabtha	I wrote it	كتَابتهَا كطبتح كتَابتهَا درت تمينَا	كتبتهَا
dert tamina	I made pasta with semolina and honey	دَارط طمن درت طمينة ضغت تمينة رحت لّ مطر	درت طمينة
ra7t lel matar	I went to Airport	غَاحط لَال متَاغ رَاحط لّ متغ رَاحت لَال متر دغط تمن	رحت لَال مطار
rahou yanzal la mtar	It rains	غح ينزَال لَ متَار رحو يَانزل ل مطاغ رهو ينزل ل مطر جَادر هد ره	رهو ينزل ل مطر
jadore had riha	I like so much this perfume	جَادور هَاد ريح جدر هد ريح جَادور هَاد ريحة جدور هَاد ريهَا	هَاد ريحة j'adore
men yatakalam?	Who is talking?	من يطكلم مَان يَاتَاكَالَام من يتَاكَلَام مَان يَاتَاكَالَام	من يتكلم

- Vowel processing: The vowels (a, i, o, u, e, y) can be replaced by different Arabic letters or by none. The last case indicates that the Arabizi letter does not translate into another Arabic letter. Vowel transliteration depends on the location of the vowel in the word. If the word begins with a vowel, the letter will be replaced in most cases by the letters إ,أ, ا . For example, the word 'oumba3d' that means "after" is transliterated as أومبعد . If a vowel appears at the end of the word, the letter will be replaced in most cases by the letters: ا ,ة, و ,ي . For example, the word 'chaba' that means "a beautiful woman" is transliterated as شَابة ^ , but the word 'ktabha' that means "he wrote" is transliterated as كتبهَا . Hence, even if these two words (chaba, ktabha) end with the same letter (a), this letter is not replaced by the same Arabic letter. It will be replaced by ة in the case of the first word and by ا in the case of the second word. A major challenge arises when these vowels appear in the middle of a word because these letters could be replaced by the letters ا ,و ,ي . In some cases these letters may not be replaced by other. For example, the word 'yal3ab' that means "it plays" or "he plays" is transliterated as يلعب . In this case, we observe that the two vowels (a) are not replaced by letters.
- The ambiguity between several letters: An Arabizi letter can correspond to several Arabic letters. For example, the letter 't' can correspond to two Arabic letters: ت is used in the expression 't3ayi' and it means "you are tiring" (تعيّ) and ط is used in the word 'tamina' and it means "toasted semolina with honey" (طمينة). Another example corresponds to the combination of the letters 'kh' that can be transliterated as a group to a single Arabic letter (خ). For example, the word 'khradj' that means "to go out" can be transliterated as خرج . However, there is another alternative in which both letters كه are transliterated separately. The above is because the letter 'h' could correspond to two Arabic letters (ه and ح). These ambiguities can be detected if we consider the set of letters that constitutes each word as a whole instead of addressing the transliteration process letter by letter.
- The ambiguity related to the context: In some cases, several transliterations can satisfy the same word. For example, the word 'matar' can be transliterated as مطر meaning "rain" but it can also be transliterated as مطار meaning "airport". The example illustrates that both words, despite having different meanings, can be considered as feasible transliterations. In this case, the context of a word can help indicate which transliteration fits best into the whole message [19].
- Ambiguity related to code switching[1]: Most Arabic speakers use two or more languages in addition to their native language. For example, Egyptian speakers use Arabic and English, and Algerians use Arabic and French. Furthermore, Arabian speakers use an Arabic dialect. For example, Egyptian speakers make use of the Egyptian dialect. Hence, it is common that messages written in Arabizi can com-

---

[1]Code switching: The presence of different language and dialects into the same message.

bine two or more languages in a single message. Thus, depending on the context, a word can have different meanings. For example, the word 'men' in English can be transliterated into Arabic as من switching to the meaning 'who'. Another example in French is the word 'j'adore' often written without apostrophe, and 'jadore' that can be transliterated as جادور meaning a very offensive person in 'Maghrebi' dialect.

3   The Research Work Inspiring Our Approach

3.1   Arabizi Transliteration

Our proposal for transliteration is inspired by several works in the research literature. For Arabizi transliteration, van der Wees et al. [29] used a table extracted from Wikipedia[2] composed by sentences written in Arabizi and Arabic. We will also use a table with sentences in Arabizi and Arabic. However, we will include a set of rules to handle the position of a letter in a word and some additional cases that were revised in Sect. 2. To introduce our table of sentences and our set of rules, we rely in our prior work [14]. The difference between our work and the work proposed by van der Wees et al. [29] can be summarised in the following two points: (1) Vowel processing: We handle the ambiguity of the vowels considering that each vowel can be replaced by different letters or by a NULL character, and (2) We handle the ambiguity of some characters that have the same sound or similar sounds. For example, the letters 's' and 'c' have similar sounds and then they can be replaced by the letters س and ص.

Our proposed transliteration approach is also inspired by the work presented in [2, 16, 19, 29–31]. All these works produce a set of feasible candidates for the transliteration of an Arabizi word into Arabic. For example, some of the feasible candidates generated using lexical variants using a letter by letter approach for the word *3afsa* can be: عافزة or عافزَى, عافزَا, عافسَا. We note that by applying this approach we could never obtain the word عفسة that corresponds to a correct transliteration in some contexts. This is mainly due to the same reason that we cited above: lexical variants using a letter by letter approach tend to replace Arabizi letters by Arabic letters. However in some context the correct meaning is pointed out by replacing the vowel with a NULL character. The originality of our work at this stage is that our algorithm is able to create all the feasible candidates.

At last, another inspiration for our proposal is taken from the work presented by [15] and also by our prior work [16, 32]. All these works make use of a statistical language model to determine the best candidates in Arabic for a given Arabizi word. On the other hand, these works assimilate the task of transliteration to a translation task, splitting each sentence into a set of words and then each word into a set of characters, conducting a kind of translation at character level. The major drawback of

---

[2]https://en.wikipedia.org/wiki/Arabic_chat_alphabet.

these approaches is that they depend on a corpus that comprises a set of transliterated messages from Arabizi to Arabic. The creation of a corpus of this type depends on humans and then it is a very time and effort consuming task.

3.2 Arabizi Machine Translation

In the context of machine translation, we take some inspiration from the work authored by May et al. [30] and van der Wees et al. [29]. May et al. [30] used a phrase-based machine translator system similar to the one proposed by Moses [38]. Their system was trained using a collection of Arabic-English sentences comprising 1.75M lines (52.9M Arabic tokens). In addition, these authors used a 5-gram English language model. The translation results between Arabic and English are up to 9.89 according to the BLEU score. van der Wees et al. [29] built a machine translation system based on an unsupervised alignment of words. The best results achieved using this approach were up to 18.4 according to the BLEU score. Our proposal is also inspired by our prior work [31, 32] where we introduced a translation system between the Algerian dialect and Arabic. We proposed a comparison between statistical translators and neural translators after a transliteration step. We showed that the quality of a transliteration process directly affects the quality of the translation. In that works we achieved a BLEU score up to 6.01 using for automatic transliteration and up to 10.74 using manual transliteration. To illustrate the utility of a transliteration process, we also experimented with translation systems without using the transliteration step. We found that the BLEU score in this last case was very low decreasing up to 4.26.

Main differences between this work and the previous work authored by May et al. [30] and van der Wees et al. [29] are the following. Whilst previous work rely on an existing parallel corpora between Englishand Arabic to apply a statistical machine translation system, we focus on Arabizi and especially those variants generated from dialects because it is a more challenging and unexplored task. In this case, the translation considers modern standard Arabic (MSA) as a pivot as we start transliterating Arabizi sentences into MSA. In our prior work [31, 32] we started working on Algerian Arabizi and then we constructed the Arabizi side of the corpus [20]. In contrast to these works, now we construct a parallel corpus between Arabizi and MSA. Afterwards, we propose a transliteration step to automatically transliterate the Arabizi side of our corpus.

3.3 Arabizi Sentiment Analysis

To the best of our knowledge, limited work has been conducted on Arabizi sentiment analysis [33] or on Arabic and Arabizi at the same time [34, 35]. In [33], the authors present a transliteration step before proceeding to the sentiment classification. However their approach presents two majors drawbacks: (1) They rely on a very basic table to transform Arabizi to Arabic, neglecting the processing of Arabizi ambiguities; (2) The authors construct a small size manually annotated corpus, containing only 3026 messages. In [34], the authors automatically construct SentiAlg, an annotated sentiment corpus dedicated to the Algerian dialect. In [35] the authors manually constructed TSAC, an annotated sentiment corpus dedicated to the Tunisian dialect. Both works make use of Doc2vec for feature extraction. Both works handled Arabizi without calling a transliteration process, fact that explains the low obtained results in classification (up to 68% for [34] and up to 78% for [35] in terms of accuracy).

Our proposed sentiment analysis approach is firstly inspired by the work discussed by [39–42] that make use of Word2vec/Doc2vec for feature extraction. However, our approach also uses fasttext for validation purposes. It is also inspired by the works authored by Dahou et al. [43] and Attia et al. [44], making use of deep learning techniques. However prior work is focused only on the use of CNN methods whilst our approach also explores the performance of LSTM, Bi-LSTM and MLP-based models. Our proposal is also inspired by the works authored by Guellil et al. [34] and Medhaffar et al. [35], what are the only works focused on Arabic and Arabizi at the same time. We particularly extend our prior work [34] presenting the automatic construction of an annotated sentiment corpus. However, in contrast to this prior work, now we use a large corpus.

4   Methodology

4.1   Arabizi Transliteration Approach

The first purpose behind our approach is to transliterate an Arabizi message to Arabic. To do that we first recover a large Arabic corpus to generate an Arabic statistical language model. Afterwards, we propose a set of rules to transliterate messages from Arabizi to Arabic. Then we apply these rules on each Arabizi word producing a set of feasible candidates. Finally, we retrieve the best candidate according to the language model. Hence, the proposed approach consists of two important phases: (1) Language model generation (presented in Fig. 1), and (2) Transliteration approach (presented in Fig. 2).

In this work we generate a statistical language model based on the frequency of each term in the corpus. This model relies on a large scale Arabic corpus extracted from social media. The extraction and processing of this corpus is discussed in more details in Sect. 4.1.1. The transliteration approach is based on three main steps which are: (1) Checking of a set of the rules for the Algerian Arabizi, (2) Generation of a list of feasible candidates, and (3) Retrieval of the best candidate based on the generated language model.

Fig. 1 High-level
architecture of our Arabizi
language model generation
process

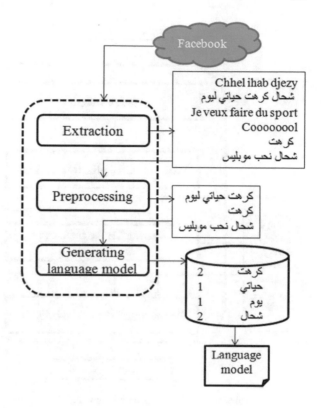

4.1.1 Creation of the Arabic Corpus and Transliteration of Arabizi Messages

Our approach begins by recovering messages from social media. The created corpus contains a set of posts and comments of Algerian speakers retrieved from the most popular Facebook pages in Algeria. However, Algerian people (like other Arabic people) switch code between MSA and dialect (Algerian dialect, in the case of Algeria). However, no need to an identification step in order to proceed to transliteration, because when people are writing in Arabizi, they also switch code between MSA Arabizi and dialect Arabizi.

After recovering this corpus, we focused on the messages written in Arabic. First, we replaced the different Arabic characters by their Unicodes. The reason for this replacement is that the same Arabic character can cope with different writing possibilities depending on its position within the word. For example the character ب can be written in four different ways. Consequently, we replaced the different variants of this letter (ب) by a single Unicode U+0628 that represents the four variants of the same letter. Then, we tokenized the corpus determining the size of the vocabulary and counting the occurrences of each vocabulary entry in the corpus.

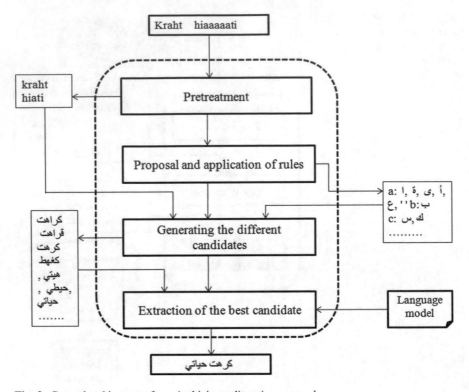

Fig. 2 General architecture of our Arabizi transliteration approach

Finally, we generated a statistical language model based on maximum likelihood estimation (MLE).

To transliterate each message, we start by turning all the characters into lowercase letters. Then we replaced some characters by characters having the same meaning in Arabizi (for example, é is replaced by i and è by e). Then, we payed attention to verbal exaggeration. Using regular expressions we reduced each exaggerated word into its canonical form (for example, 'mliiiiiiiiiiih' was transformed into 'mlih' and فوّوّوّوّوّور was transformed into فور). We also removed blanks at the beginning and at the end of each message as well as the successive blanks.

4.1.2 Transliteration Rules

Each Arabizi letter could be replaced not only by a single Arabic letter but with many. For example the letter 'a' can be replaced by six different Arabic letters as:

- a = ' ا '. In the word *babour*, that means "boat", the transliteration is ' بَابور '.
- a = ' ة '. In the word *chaba*, that means "beautiful", the transliteration is ' شَابة '.

- a = ' ى '. In the word *3la*, that means "on", the transliteration is ' عَلَى '.
- a = ' أ '. In the word *akhdam*, that means "work", the transliteration is ' أخدم '.
- a = ' ع '. In the word *asas*, that means "guardian", the transliteration is ' عَسَاس '.
- a = ' ' (NULL). In the word *3afsa*, that means "trick", the transliteration is ' عفسة '.

In Table 2, different possibilities to transform letters from Arabizi to Arabic are presented.

Based on Table 2 and on Arabizi transliteration analysis, a set of rules for transliteration from Arabizi to Arabic is proposed. The rules are:

1. At the beginning of the word, the vowel "a" is replaced by the letter " أ ". In the middle of the word, this letter can be replaced either by the Arabic character ' ا ' or by the empty character (NULL). At the end of the word, in addition to the two characters ' ا ' and the empty character, the 'a' can also be replaced by the letter ' ة ' or ' ى '. The same rationale is also applied for other vowels (i.e. replacing them according to their positions in words).
2. When a word starts with 'el' or 'al', we directly replace these two letters by ' ال '.

Table 2 Some Arabizi letters have many possibilities to be transliterated to Arabic

Arabizi letter	Arabic letter	Arabizi letter	Arabic letter
a	ا, ة, ى, أ, ع, "	p	ب
b	ب	q	ك
c	س , ك	r	غ ,ر
d	ظ, ض, د	s	ص, س
e	ا, "	t	ط, ت
f	ف	u	و, أ, "
g	ق	v	ف
h	ح, ه	w	و
i	ي, "	x	كس
j	ج	y	ي, ا, "
k	ق, ك	z	ز
l	ل	7	ح
m	م	5	خ
n	ن	3	ع
o	و, أ, "	9	ق

3. The combination of the two letters 'ch' can be interpreted either by: ' سه ', ' که ',
   ' سّح ', ' خّ ' and then by applying the different substitutions depicted in Table 1.
   However, the merge of these two letters can also be interpreted as the letter ' ش
   '. We notice the same phenomenon for the letters 'dh', 'th' and 'kh' which, in
   addition to the substitutions appearing in Table 1, can also be interpreted as: '
   ض ', ' ث ' and ' خّ '.

4.1.3   Generation of Candidates

By applying different letter replacement options and rules, each Arabizi word gives
birth to several word candidates in Arabic. For example, the word "kraht" gives 32
possible candidates as قَراهت ,كَراهت and كرهت , among others. For example,
the letter "k" can be replaced by the Arabic letters ك and ق . The letter "r" can
be replaced by the letters ر and غ . The letter "a" can be replaced by six letters,
however, as it appears in the middle of the word it can only be replaced by two
feasible possibilities: i.e. ا and the NULL (' ') character. The letter "h" could be
replaced by the Arabic letters: ه and ح . Finally, the letter "t" could be replaced by
the letters: ت and ط . Hence, the number of transliterations of the word "kraht"
are equal to 2^5, i.e. this word has 32 candidates. However, just one transliterated
word for "kraht" is correct: " كرهت ". Another example corresponds to the word
"hiati" that gives 16 candidates as هيَاتي and هيتي , among others. However, the
right transliterated word is only one: " حيَاتي ".

4.1.4   Best Candidate Selection

To determine the best candidate for the transliteration of a given Arabizi word into
Arabic, we use a statistical language model. The idea is to recover for each candidate
its probability of occurrence inferred from the language model and finally, to select the
candidate with the highest probability. In the running example for the words "kraht"
and "hiati", the best candidates in terms of number of occurrences are " كرهت " and
" حيَاتي ". A great proportion of the other generated candidates obtain a probability
close to 0 because they are not used in Arabic or its dialects. Some other candidates
could obtain a greater probability but in almost all cases, it will be not significant
compared to the best candidate. In the case that none of the words generated belong
to the vocabulary, we will consider that the best candidate corresponds to the word
that contains the Arabic letters in the order indicated by the letters shown in Table 2.
Note that in Table 2, we sorted Arabic letters according to the number of occurrences
in our corpus. In our example, if both words "kraht" and "hyati" do not belong to

the vocabulary of our language model, we will consider that the best candidates are كَرَهت and هيَاتي , respectively. In the first case, the selected candidate matches with the actual transliterated Arabic word but in the second case this rule does not match the right transliteration.

The proposed transliteration approach is simple. Our rule-based approach make use of a language model in order to select the best transliterated candidate. A number of research works in the literature has proposed statistical or neural-based method for transliteration. However, these approaches show good results depending on corpus size and coverage. As such Arabizi corpus does not exist, our approach based on rules is feasible and practical. Afterwards, our approach can be useful to automatically build an Arabizi corpus.

4.2 Application of Transliteration to Translation and Sentiment Analysis

4.2.1 Application of Transliteration to Automatic Translation

In this work, a number of text messages written in Arabic were extracted from Facebook/Twitter. To extract these posts, we used RestFB[3] and Twitter4j[4] APIs. These posts were extracted from the most popular pages used in Arabic countries as MustafaHosny,[5] that records 32,854,861 fans in Egypt, ooredooqatar,[6] that records 834,031 fans in Qatar, or EnnaharTv,[7] that records 9,603,348 fans in Algeria. The most popular Facebook pages were identified using statistics offered by SocialBakers.[8] Using these methods, the collected corpus achieved a huge size. This corpus contains Arabic comments written in both scripts: Arabic and Arabizi. An initial treatment was done to separate both parts of the corpus (Arabic and Arabizi).

Inspired by the work presented in [20, 45, 46] on statistical machine translation of Arabic and its dialects, we propose three main steps for translation: (1) Language model training. (2) Word-alignment. (3) Model tuning. To train the language model we used the part of our extracted corpus that was written in Arabic. The objective of the word alignment task is to discover the word-to-word correspondences in a sentence pair. Tuning refers to the process of finding the optimal weights for the translation model. Optimal weights are those which maximise translation performance on a small set of parallel sentences (the tuning set). Translation performance

---

[3] http://restfb.com/.
[4] http://twitter4j.org/en/index.html.
[5] https://www.facebook.com/MustafaHosny/.
[6] ooredooqatar.
[7] https://www.facebook.com/EnnaharTv/?_ga=2.64384426.442009333.1538003328-860660030. 1538003328.
[8] https://www.socialbakers.com/statistics/facebook/pages/total/algeria/.

is usually measured with BLEU score, but the tuning algorithms all support (at least in principle) the use of other performance measures. Subsequently alignment model and tuning methods were used to select the best translation. More details about the language, alignment and tuning models that we use are in given in Sect. 5.1.2.

4.2.2 Application to Sentiment Analysis

For sentiment analysis, we followed three main steps: (1) Automatic construction of an Arabic sentiment lexicon. (2) Automatic annotation of Arabic messages, and (3) Sentiment classification of Arabic messages. As almost all the sentiment lexicons which were constructed are dedicated to MSA or the most studied dialect such as Egyptian or Iraqi dialects, we firstly propose to automatically construct a sentiment lexicon covering MSA and all the dialects. In this work, our approach was applied to MSA and Algerian dialect to facilitate some aspects of the study, such as the validation of the results achieved (for further details on this point, please refer to Sect. 5.1.1).

The sentiment lexicon was constructed by translating an existing English lexicon, namely SOCAL [47] to Arabic using the Glosbe API.[9] After the automatic translation, the same score is assigned to all the translated words. This score corresponds to the score of English word from which they are translated. For example, all the translations of the English word 'excellent' with a score of +5, such as باهي (bAhy), لطيف (lTyf), and ' مليح ' (mlyH), are assigned a score of +5. The different obtained terms were tagged from negative(labels ranging between −1 and −5) to positive (labels ranging between +1 and +5). Since some Arabic sentiment words result from different English words having different sentiment scores, an average score is assigned to such English words. For example, the word ' مليح ' obtained the score (+5) when it is associated to the English term 'excellent', however, the same term obtained the score (+3) when it is associated to "good". Hence, an average of all sentiment where the term مليح is associated is then calculated. Finally the resulted lexicon should be manually reviewed.

The constructed lexicon is used to automatically provide a sentiment score for Arabic message utterances. Doing so provides a method to automatically construct a large sentiment training corpus. To calculate the score, different steps are followed:

1. Opposition: Opposition is generally expressed with a word. For example in Algerian dialect, it corresponds to the word بصح (bSH—but). For handling opposition, we only take into consideration the part of the message following ' بصح '. For example, only the part in bold in the message: زاني مريضة حبيت نخرج نلعب نفرح بصح —I wanted to go out to play to be happy but **I'm sick**) is considered leading to a negative sentiment.

---

[9]https://glosbe.com/en/arq/excellent.

2. Multi-word expressions: Since our lexicon contains multi-word entries, we extract all single and multi-word expressions appearing in our lexicon.
3. Handling Arabic morphology: we employ a simple rule-based light stemmer that handles prefixes and suffixes. Since words typically are composed of prefix(es)+stem+suffix(es), if a word does not match any of the entries in our lexicon, we try to remove all possible prefix/suffix combination to see if the remaining possible stems would match entries in our lexicon. For example, the word بكتابو (and with his book), which is optimally segmented as و+ب+كتاب+و (w+b+ktAb+w), the stemmer would produce the following possible forms: كتابو ,بكتاب ,بكتابو ,وبكتاب and كتاب . The prefixes that we used are: و (w), ا (A), ي (y), ت (t), ن (n), ب (b), and ال (Al) . As for the suffixes, we used: ي (y), ت (t), و (w), ا (A), ة (p), ين (yn), ا (A), ه (h), هم (hm), كم (km), نا (nA), ها (hA), هو (hw), ك (k), ني (ny), لهم (lhm), لكم (lkm), نا (nA), لنا (lnA), لها (lhA), لو (lw), لك (lk) لي (ly). However, this list could easily be extended. We handle negation prefixes and suffixes separately. Since some words may end with ى (Y) (ex. بكى (bkY—cried)) when they are in isolation, but turn into ي (y) when suffixes are attached (ex. بكيت (I cried)), we normalize ى to ي .
4. Negation: negation can reverse polarity. Negation in Arabic and its dialects is usually expressed as an attached prefix, suffix, or a combination of both. For example, the word مانحبكم ش (mAnHbkm$—I don't like you) can be written in the following way: مَا نحبكم ش (mA nHbkm$), مَانحبكم ش (mAnHbkm $), or مَا نحبكم ش (mA nHbkm $). Therefore, we notice that negation can be attached to or separated from words. We deal in this work with agglutinated and separated negation markers. We define a list of prefixes and suffixes related to negation. We have found that in most cases, negation does affect not only the preceding word but also some of the words in the rest of the sentence. Thus, once a prefix or negation suffix is detected, we reverse the score of the words succeeding the negation (multiplying the score by (-1)).

To increase labelling precision, we used the following heuristics:

1. if there are more positive sentiment words than negative sentiment words, then the message is considered positive (and vice versa).
2. if the number of positive and negative sentiment words is equal, then we don't label the message.
3. the number of positive/negative of sentiment words had to constitute at least 25% of the words in the message.

For classification, we used shallow and deep models of classification. For both classification approaches, we extracted features using word embedding techniques. For shallow classification we used Word2vec and Doc2vec. For deep classifiers we used an embedding layer and fastText. For word2vec, we used both Skip-Grams (SG) and Continuous Bag-Of-Words (CBOW) [48]. In addition, for Doc2vec we used both architectures namely distributed memory paragraph vectors (PV-DM) and distributed bag-of-words paragraph vectors (PV-DBOW) [49]. We used different machine learning algorithms for classification such as support vector machines (SVM) and naive Bayes (NB). For deep classification we used CNN, LSTM and MLP-based neural networks.

5 Experiments and Results

5.1 Experimental Setup

In this section we present the experiments that we carry out to study our proposal. First of all we present the results obtained from our transliteration process. Then we apply this approach on different Arabizi datasets and present the different results related to automatic translation and sentiment analysis considering two specific tasks: automatic translation and sentiment analysis. To study how useful is the transliteration process in these tasks we evaluate the results before and after the transliteration process. At the end, our aim is to answer to the question that motivates this study: **Is transliteration crucial for automatic translation and sentiment analysis?** Before presenting our results, we introduce the general experimental setup related to this study.

5.1.1 Data

To validate our transliteration approach, we study its performance in three datasets. The first one is named Corpus_50 and is a part of the Cottrell's corpus [4] used in [16, 31, 32]. This corpus contains 50 messages written in Arabizi along with their manual transliteration. The second corpus is named Corpus_200 and it was constructed and transliterated in the context of the work of [16, 31, 32]. This corpus corresponds to a part of the PADIC corpus [20] that is focused on the translation of Algerian to MSA. However this corpus was not initially written in Arabizi but in Arabic. Hence, the authors transliterated this corpus from Arabic to Arabizi in a semi-automatic way, applying a set of rules for automatic transliteration from Arabic to Arabizi. At last, the transliterated sentences were manually reviewed by human native speakers. The third corpus is named Senti_Alg_test and is a test corpus created and manually annotated by Guellil et al. [34]. This corpus contains 1000 messages in Algerian dialect, of which 500 are written in Arabic and the other 500 are written in Arabizi. In the context of this study we focus on the Arabizi part. We transliterated this corpus automatically based on the transliteration approach used to obtain Senti_Alg_test_Trauto, that is the automatic transliterated version of Senti_Alg_test.

To validate our transliteration approach we manually translated a dataset namely Arabizi_translate composed by a collection of posts extracted from two social media sources (Facebook and Twitter). This dataset contains 2924 messages of which 80% was used in training and 20% in validation. To study the performance of our approach, we used two datasets. The first one was Senti_Alg_test used for sentiment analysis tasks. The second one was Corpus_200 that was manually translated during the creation of PADIC [20]. Hence, we rely on this corpus that was semi-automatically transliterated in Arabizi by Guellil et al. [16, 31, 32].

For the sentiment annotation process, we rely on sentiment lexicon constructed by Guellil et al. [36] and used by the same authors in [34]. This lexicon contains 2384 entries. In the context of this study, we manually reviewed this lexicon to obtain a new dataset named Senti_lex that contains 1745 terms of which 968 are negative, 6 are neutral and 771 are positive. As we focus on a binary classification between positive and negative classes, the unbalance between neutral an positive/negative classes is not crucial. Moreover, many other lexicons presented in the literature are unbalanced (see for instance [50], a lexicon that comprises 3982 entries of which 856 words are positive, 636 words are negative and 2490 words are neutral).

Two large corpus (Ar_corpus1) and (Ar_corpus2) were automatically extracted from Facebook. The first one (Ar_corpus1) focus on 133 famous Arabic pages as *Oreedo*, *Djeezy*, and *Hamoud Boualem*. This dataset was collected during September of the year 2017 and it contains 8,673,285 messages of which 3,668,575 are written in Arabic. The second dataset (Ar_corpus2) was collected during November of the year 2017 and it is focused on 226 famous Arabic pages. This dataset contains 15,407,910 messages of which 7,926,504 are written in Arabic. Ar_corpus1 was used in the transliteration process and Ar_corpus2 was used in sentiment analysis. Using Ar_corpus2 and Senti_lex we constructed ALG_Senti, an automatically annotated corpus that contains 255,008 messages of which 127,004 are positive and 127,004 are negative. Senti_Alg_test was used to validate our sentiment analysis approach. The second dataset that we used to validate our sentiment analysis approach was TSAC,[10] a Tunisian sentiment corpus that combines Arabic and Arabizi messages that was manually annotated by Medhaffar et al. [35]. This corpus contains a partition of 13,669 messages used for training (TSAC_Train) of which 7154 messages are positive and 6515 are negative. A second partition of the dataset was used during the testing step (TSAC_test) and it contains 3400 messages of which 1700 messages are positive and 1700 are negative.

5.1.2 Models

To study the performance on machine translation we followed the approach exposed in [51]. Therefore, we used the open-source Moses toolkit [38] to build a phrase-

---

[10]https://github.com/fbougares/TSAC.

based machine translation system using bidirectional phrase modelling, lexical translation and a tri-gram language model. We used GIZA++ [52] for alignment and KenLM [53] to compute the tri-gram language model. For the evaluation of our machine transliteration system we used accuracy and the BLEU metric [54].

For classification tasks we used different models. We studied the performance of models based on Word2vec/Doc2vec representations, classical machine learning algorithms and embedding-based layers as fasText. We also studied the performance of different deep learning representation methods. For Word2vec/Doc2vec models, we used the Gensim toolkit.[11, 12] For both representations we used a context window of 10 words to produce representations using CBOW, SG, PV-DBOW and PV-DM, all of them with a dimensionality equal to 300. We trained the Word2vec/doc2vec representations on the messages that belongs to training data partitions. For the classification model we used the implementation developed by [40] that make use of Word2vec to ingest 5 classifiers: GaussianNB (GNB), LogisticRegression (LR), RandomForest (RF), SGDClassifier (SGD, with loss = 'log' and penalty = 'l1') and LinearSVC (LSVC with C = '1e01'). We also used the representation model studied in [34] named Doc2vec.

To study the performance of neural networks-based algorithms we used the approach presented in [44]. Therefore, four models were evaluated: CNN, LSTM, MLP, and Bi-LSTM. These models were implemented using embedding functions provided in *Keras*.[13] For training this layer, we use a default sitting (with input_dim = 139,297 for ALG_Senti and 21,937 for TSAC_train and output_dim = 300). In Table 3 we provide more details about the configuration of each architecture with the output shape and the number of parameters for each training set (ALG_Senti and TSAC_train). For all the used models we considered 100 epochs with an early_stopping parameter to avoid iterations in the absence of improvements. This configuration allows us to stop the models on average around the 20 epochs. For all our models, we used the Adam optimiser. For sentiment classification, three metrics were used: Precision (P), Recall (R) and F1-Score (F1).

5.2   Experimental Results

5.2.1   Transliteration Performance

Our transliteration system was evaluated on three data sets: (1) Corpus_50, (2) Corpus_200, and (3) Senti_Alg_test. Corpus_50 was manually transliterated to obtain Corpus_50_Trmanu. This corpus was used in [16, 31, 32]. Corpus_200 was semi-automatically transliterated to obtain Corpus_200_tr [16, 31, 32]. Senti_Alg_test was also manually transliterated to obtain the corpus Senti_Alg_test_trmanu. One

---

[11] https://radimrehurek.com/gensim/models/word2vec.html.

[12] https://radimrehurek.com/gensim/models/doc2vec.html.

[13] https://keras.io/layers/embeddings/.

Table 3 Neural network architectures studied in our work

Model	Layers	ALG_Senti		TSAC_train	
		Output shape	Params	Output shape	Params
CNN	embedding_1(default)	(None, 12,300)	41789100	(None, 16,300)	6581100
	conv1d_d	(None, 12,300)	630300	(None, 16,300)	630300
	global_max_pooling1d_1	(None, 300)	0	(None, 300)	0
	dropout_1	(None, 300)	0	None, 300)	0
	dense_1	(None, 600)	180600	(None, 600)	180600
	dense_2	(None, 2)	1202	(None, 2)	1202
MLP	embedding_1(default)	(None, 12,300)	41789100	(None, 16,300)	6581100
	dense_1	(None, 12, 64)	19264	(None, 16, 64)	19264
	global_max_pooling1d_1	(None, 64)	0	(None, 64)	0
	dropout_1	(None, 64)	0	(None, 64)	0
	dense_3	(None, 600)	39000	(None, 600)	39000
	dense_4	(None, 2)	1202	(None, 2)	1202
LSTM	embedding_1(default)	(None, 12,300)	41789100	(None, 16,300)	6581100
	lstm_1	(None, 16, 64)	93440	(None, 12, 64)	93440
	global_max_pooling1d_1	(None, 64)	0	(None, 64)	0
	dropout_1	(None, 64)	0	(None, 64)	0
	dense_1	(None, 600)	39000	(None, 600)	39000
	dense_2	(None, 2)	1202	(None, 2)	1202
Bi-LSTM	embedding_1(default)	(None, 12,300)	41789100	(None, 16,300)	6581100
	bidirectional_1	(None, 12,128)	186880	(None, 16,128)	186880
	global_max_pooling1d_1	(None, 128)	0	(None, 128)	0
	dropout_1	(None, 128)	0 (None, 128)	0	
	dense_1	(None, 600)	77400	(None, 600)	77400
	dense_2	(None, 2)	1202	(None, 2)	1202

Table 4 Accuracy transliteration results

Corpus	1%	5%	10%	25%	50%	75%	100%
Corpus_50	67.36	70.02	72.49	73.24	74.19	74.57	**74.76**
Corpus_200	67.31	69.74	69.67	70.95	71.06	71.28	**72.03**
Senti_Alg_test	66.17	69.17	70.15	70.91	71.23	71.60	**72.05**

important aspect in our transliteration approach was the use of a large corpus written in Arabic letter (Ar_corpus1) to generate the language model. To show the importance of this large dataset, we separated it into 7 different partitions containing 1, 5, 10, 25, 50, 75, and 100% of the Ar_corpus1. The effect of corpus size on transliteration quality (accuracy) is presented in Table 4.

It can be seen in Table 4 that the size of the corpus used to create the language model directly affects the performance of the transliteration system. As the table shows, the performance gradually improves as the corpus increases in size. On Corpus_50, the results reach 67.36% when using 1% of corpus while the performance rises to 74.76% when the entire corpus is used, achieving a real gain equivalent to 7.4%. The same observation can be done on Senti_Alg_test with a real gain equivalent to 5.88%. These results are summarised and compared with other methods in Sect. 6.

5.2.2   Machine Translation Results

Our translation system was evaluated on two datasets: (1) Senti_Alg_test, which was manually translated in the context of this study, and (2) Corpus_200, which was a part of the PADIC corpus and it was manually translated by Arabic native speakers. Both datasets were automatically transliterated using our system. To illustrate the usefulness of our transliteration system in the translation process, we carry out two kind of experiments: (1) Direct translation from the Arabizi test corpus to MSA, and (2) Translation from a transliterated test corpus to MSA. In the first experiment we considered the Arabizi part of Senti_Alg_test that we directly translate to MSA. In the second experiment, we firstly transliterated the training partition and then we used the testing part of both test corpus to study the performance of our proposal (i.e. Senti_Alg_test and corpus_200). Table 5 presents the BLEU scores obtained in both test datasets.

It can be seen in Table 5 that transliteration improves the performance in terms of the BLEU score. While the results obtained are 8.13 and 0.0 when the transliteration step is not used, these results improve rising up to 13.06 and 8.35 after using the transliteration process. Another interesting observation is related to the BLEU score achieved in the Corpus_200 when the transliteration step is not used. As Table 5 shows, for this configuration the BLEU score is 0.0. This result is explained as follows. Although this corpus (Corpus_200) shares some words with the vocabulary of the training dataset, many of these words can not be recognised due to the presence of Arabizi passages. These results show that by not using a transliteration process the translation process can completely fail.

Table 5 Results of machine translation on Senti_alg_test and Corpus_200

Experiment	Test set	BLEU score
Without transliteration	Senti_Alg_test	8.13
	Corpus_200	0.00
With transliteration	Senti_Alg_Test	**13.06**
	Corpus_200	**8.35**

5.2.3 Sentiment Analysis Results

Our sentiment classification system was evaluated in two datasets: (1) Senti_Alg_test, and (2) TSAC_test. The Arabizi part of both corpus was transliterated to Arabic using our transliteration system. After the messages were transliterated to Arabic, they were ingested into the classifiers to study the performance in sentiment analysis. Our automatically annotated sentiment corpus (ALG_Senti) was used for the classification of Senti_Alg_test and TSAC_Train was used for the classification of TSAC_test. Tables 6 and 7 present the results obtained in sentiment analysis. The best results in terms of the F1 measure are depicted using bold fonts.

Table 6 Shallow classification results

Representation	Type	ML Alg	Senti_Alg_test			TSAC_test		
			P	R	F1	P	R	F1
Word2vec	CBOW	GNB	0.75	0.38	0.50	0.50	0.98	0.66
		LR	0.74	0.56	0.64	0.51	0.99	0.68
		RF	0.65	0.52	0.57	0.75	0.67	0.71
		SGD	0.69	0.62	0.66	0.84	0.11	0.20
		LSVC	0.76	0.55	0.64	0.57	0.82	0.68
	SG	GNB	0.80	0.51	0.62	0.51	0.99	0.67
		LR	0.74	0.70	**0.72**	0.64	0.82	0.72
		RF	0.72	0.60	0.65	0.84	0.77	0.80
		SGD	0.75	0.67	0.71	0.64	0.83	0.72
		LSVC	0.73	0.71	**0.72**	0.80	0.84	0.82
Doc2vec	PV-DM	GNB	0.69	0.68	0.68	0.68	0.66	0.64
		LR	0.70	0.69	0.68	0.86	0.86	0.86
		RF	0.66	0.66	0.66	0.85	0.85	0.85
		SGD	0.72	0.71	0.70	0.86	0.86	0.86
		LSVC	0.69	0.68	0.68	0.88	0.88	0.88
	PV-DBOW	GNB	0.74	0.73	**0.73**	0.72	0.70	0.70
		LR	0.71	0.70	0.70	0.90	0.90	0.90
		RF	0.70	0.70	0.70	0.89	0.89	0.89
		SGD	0.71	0.70	0.70	0.90	0.90	0.90
		LSVC	0.71	0.70	0.69	0.91	0.91	**0.91**

Table 7 Deep learning classification results

Model	Type	MDL Algo	Senti_Alg_test			TSAC_test		
			P	R	F1	P	R	F1
Keras embedding layer		CNN	0.67	0.67	0.67	0.88	0.88	0.88
		MLP	0.64	0.63	0.63	0.93	0.93	**0.92**
		LSTM	0.68	0.68	0.68	0.90	0.90	0.90
		Bi-LSTM	0.70	0.69	0.68	0.90	0.90	0.90
FasText	CBOW	CNN	0.69	0.69	0.69	0.72	0.70	0.69
		MLP	0.70	0.70	0.70	0.67	0.61	0.57
		LSTM	0.71	0.71	**0.71**	0.71	0.70	0.70
		Bi-LSTM	0.68	0.68	0.68	0.74	0.73	0.73
	SG	CNN	0.68	0.68	0.68	0.86	0.86	0.86
		MLP	0.69	0.69	0.69	0.84	0.84	0.84
		LSTM	0.69	0.69	0.69	0.85	0.85	0.85
		Bi-LSTM	0.71	0.70	0.70	0.86	0.86	0.86

It can be seen in Table 6 that SG and PV-DBOW give better results than CBOW and PV-DM in both datasets. With regard to the classifiers, LSVC and LR give its best results with Word2vec on Senti_Alg_test, whilst GNB gives its best result with Doc2vec. LSVC also achieves good results on TSAC_test with both representation models (Word2vec and Doc2vec).

With regard to neural networks-based classifiers (see results in Table 7), the best result on Senti_Alg_test was achieved using CBOW implemented with fasText. It can be seen in Table 7 that MLP performs well in TSAC_test. In addition, both tables show that the results obtained in TSAC_test are consistently better than those obtained in Senti_Alg_test. This observation is mainly due to the fact that TSAC was manually constructed while Senti_Alg was automatically constructed.

6   Discussions and Analysis

Table 8 summarises the best results presented in Tables 4, 5, 6 and 7 for transliteration, machine translation and sentiment analysis tasks, respectively. For machine translation and sentiment analysis the results were obtained after applying our transliteration approach.

Table 8 Results obtained in transliteration, machine translation and sentiment analysis tasks using accuracy, BLEU score and F1 measure, respectively

Task	Corpus	Our results	Literature work	Other work's results
Transliteration	Corpus_50	**74.76**	[16]	45.35
	Corpus_200	72.03	[16]	**73.66**
	Senti_Alg_test	**72.05**	–	–
MT	Corpus_200	**8.35**	[32]	6.01
			[32]	
	Senti_Alg_test	**13.06**	–	–
SA	Senti_Alg_Test	**73.00**	[34]	68.00
	TSAC_test	**92.00**	[35]	78.00

We compare the results obtained using our proposal with the results obtained by other methods on the same datasets either by applying a different transliteration approach as in [16, 31] as well as by directly handling Arabizi (i.e. without applying the transliteration step [34, 35]).

It can be seen in Table 8 that all the results presented in this paper (except those related to transliteration of Corpus_200) outperform the results presented in the literature. For the Corpus_200, the results presented in [16] are slightly better than those presented in this work. It can also be seen that our transliteration system improves the performance achieved in machine translation and sentiment analysis tasks. The real gain in sentiment analysis is 14% in F1 measure over its competitor [35]. Despite the fact that the transliteration was not improved in Corpus_200, the BLEU score obtained after the use of our transliteration system outperformed the state of the art.

Even though our system achieves a good performance, our transliteration system has some limitations. After manually inspecting the results obtained by our transliteration system, we detected some errors in the process. We attribute these errors mainly to the technique used to select the best candidate for transliteration. Since our transliteration system is based on language models, the most promissory candidates are selected based on their frequency. However, frequentist approaches do not always drive to the best solutions. For example, the word "rakom" that means "you are" is transliterated as " رقم " meaning "a number". Even though that the correct transliteration is " راكم ", our system selects " رقم " because it is a more frequent word in the corpus. There are two factors that explain that this word is more frequent: (1) Its meaning, i.e. this word is more used with this meaning than with another, and (2) Its length, i.e. shorter words tend to be more frequent than longer words.

One possible solution to address this problem is to consider other variables that allow determining the best candidates. A variable that looks promising to address this issue is the inverse frequency of a word (idf). As well as in the traditional Tf-Idf information retrieval model, the combination of term frequency (Tf) and inverse document frequency (Idf) allows to address the trade-off between descriptive and discriminative capacities. The use of a language model that considers both aspects would allow to address this issue.

Our inspection also detected some errors in translation and in sentiment analysis attributable to errors in the transliteration process. For example, "khlwiya" that means "good" or "quit" was erroneously transliterated as " خلْيَا " meaning "empty". The correct transliteration is خلوية. Thus, an incorrect transliteration drives to an incorrect translation. Improvements in the transliteration process can drive to improvements in machine translation. Our machine translation process is based on a manually constructed corpus of relatively small size. Note that this corpus only contains 2924 messages manually translated to MSA. The BLEU score obtained by our method is mainly due to the small size of the corpus used. We believe that by increasing the size of the corpus we will obtain much better results. A possible alternative to explore in this issue relies on the use of automatic or semi-automatic techniques for corpus creation.

Improvements in the transliteration process can also drive to improvements in sentiment analysis performance. Our sentiment analysis approach is based on automatic corpus annotation. This annotation was done using a sentiment lexicon. Hence, the quality of the sentiment lexicon definitely affect the quality of the automatic annotation process. Our inspection detected some errors in the lexicon. Some of the factors that explain these errors are: (1) Use of grammar irregularities in plurals, (2) Absence of specific words in the lexicon (a.k.a. out-of-vocabulary problem), and (3) Insufficient processing of language intensifiers (e.g. adjectives).

Plurals are formed using suffixes. However, for some words some specific lemmas are completely changed making it difficult to use a rule of plurals. For example, the plural of the word مليح that means "good" is not مليحون or مليحين but it is مِلَاح. The first two alternatives can be deduced using a suffix-based rule of plurals whereas the second alternative corresponds to an irregularity of the language. In addition, some words as كَادو ("a gift") are not included in our lexicon, an issue known as the out-of-vocabulary problem. Consequently, for these words our system will be unable to infer a polarity score. Finally, some adjective words as بزَاف ("very") are used to intensify the strength of a polarity word. Then it is crucial to handle them to improve the performance of sentiment analysers.

7 Conclusion and Future Work

In this paper we have presented an analytical study about the role of Arabizi transliteration in automatic translation and sentiment analysis. In order to answer to the question: **Is transliteration crucial for machine translation and sentiment analysis?** a number of different methods and resources were developed. For Arabizi transliteration, we used a rule-based transliteration approach combined with a statistical language model constructed over a huge Arabic corpus. For machine translation, we constructed the first parallel corpus between Arabizi (Algerian dialect) and MSA, applying a statistical machine translation system. For sentiment analysis, we manually reviewed a sentiment lexicon that was automatically constructed. Based on this lexicon we automatically constructed an annotated sentiment corpus containing 255,008 messages. For sentiment classification we used either shallow and deep classifiers doing use of different representation models as Word2vec, Doc2vec and fasText. Our results show that the use of Arabizi transliteration consistently improves the performance of automatic translation and sentiment analysis.

In spite of the fact that our results have a good performance, our approach presents some limitations. We are currently working to address these challenges. Our work plan includes:

- **To propose a new statistical machine transliteration system**: We plan to use the current system to automatically transliterate a larger Arabizi corpus. Using this new corpus it will be possible to design a new automatic transliteration system.
- **To propose a new Arabizi identification module**: We plan to work on a new Arabizi identification system. We have previously proposed a bilingual lexicon [18] and a set of rules for Arabizi detection [55]. However, that approach has many drawbacks related to language ambiguities. Currently, we plan to modify our identification module, improving the design process of the language classifier.
- **To enrich our parallel corpus between Arabizi and MSA**: Our parallel corpus contains 2924 messages. We propose to merge this corpus with other parallel corpus proposed for MSA and its dialects. We also propose to use an automatic approach to enrich it.
- **To automatically enrich the proposed sentiment lexicon**: We propose to enrich the proposed lexicon using Word2vec. The idea of using Word2vec is to return the words that are semantically closest to a given word. Using this approach it will be possible to provide a lexicon of much larger size.

Acknowledgements Mr. Mendoza acknowledge funding support from the Millennium Institute for Foundational Research on Data and also by the project BASAL FB0821. The funder played no role in the design of this study.

References

1. I. Guellil, H. Saâdane, F. Azouaou, B. Gueni, D. Nouvel, Arabic natural language processing: an overview. J. King Saud Univ.-Comput. Inf. Sci. (2019)
2. K. Darwish, Arabizi detection and conversion to Arabic. arXiv preprint arXiv:1306.6755 (2013)
3. A. Bies, Z. Song, M. Maamouri, S. Grimes, H. Lee, J. Wright, S. Strassel, N. Habash, R. Eskander, O. Rambow, Transliteration of Arabizi into Arabic orthography: developing a parallel annotated Arabizi-Arabic script SMS/chat corpus, in *Proceedings of the EMNLP 2014 Workshop on Arabic Natural Language Processing (ANLP)*, 2014, pp. 93–103

4. R. Cotterell, A. Renduchintala, N. Saphra, C. Callison-Burch, An Algerian Arabic–French code-switched corpus, in *Workshop on Free/Open-Source Arabic Corpora and Corpora Processing Tools Workshop Programme*, 2014, p. 34

5. A. Abdelali, K. Darwish, N. Durrani, H. Mubarak, Farasa: a fast and furious segmenter for Arabic, in *Proceedings of the 2016 Conference of the North American Chapter of the Association for Computational Linguistics: Demonstrations*, 2016, pp. 11–16

6. A. Pasha, M. Al-Badrashiny, M.T. Diab, A. El Kholy, R. Eskander, N. Habash, M. Pooleery, O. Rambow, R. Roth, MADAMIRA: a fast, comprehensive tool for morphological analysis and disambiguation of Arabic, in *LREC*, vol. 14 (2014), pp. 1094–1101

7. S. Yousfi, S.-A. Berrani, C. Garcia, ALIF: a dataset for Arabic embedded text recognition in TV broadcast, in *2015 13th International Conference on Document Analysis and Recognition (ICDAR)* (IEEE, 2015), pp. 1221–1225

8. G. Inoue, N. Habash, Y. Matsumoto, H. Aoyama, A parallel corpus of Arabic-Japanese news articles, in *LREC* (2018)

9. S. Mohammad, M. Salameh, S. Kiritchenko, Sentiment lexicons for Arabic social media, in *LREC* (2016)

10. N. Al-Twairesh, H. Al-Khalifa, A. AlSalman, Arasenti: large-scale twitter-specific Arabic sentiment lexicons, in *Proceedings of the 54th Annual Meeting of the Association for Computational Linguistics (Volume 1: Long Papers)*, vol. 1 (2016), pp. 697–705

11. K. Darwish, H. Mubarak, A. Abdelali, M. Eldesouki, Y. Samih, R. Alharbi, M. Attia, W. Magdy, L. Kallmeyer, Multi-dialect Arabic pos tagging: a CRF approach, in *LREC* (2018)

12. N. Habash, F. Eryani, S. Khalifa, O. Rambow, D. Abdulrahim, A. Erdmann, R. Faraj, W. Zaghouani, H. Bouamor, N. Zalmout et al., Unified guidelines and resources for Arabic dialect orthography, in *LREC* (2018)

13. S. Shon, A. Ali, J. Glass, Convolutional neural networks and language embeddings for end-to-end dialect recognition. arXiv preprint arXiv:1803.04567 (2018)

14. I. Guellil, F. Azouaou, Asda: Analyseur syntaxique du dialecte alg érien dans un but d'analyse s é mantique. arXiv preprint arXiv:1707.08998 (2017)

15. K. Darwish, Arabizi detection and conversion to Arabic, in *Proceedings of the EMNLP 2014 Workshop on Arabic Natural Language Processing (ANLP)*, 2014, pp. 217–224

16. I. Guellil, F. Azouaou, M. Abbas, S. Fatiha, Arabizi transliteration of Algerian Arabic dialect into modern standard Arabic, in *Social MT 2017: First workshop on Social Media and User Generated Content Machine Translation (Co-located with EAMT 2017)*, 2017

17. N. Habash, A. Soudi, T. Buckwalter, On Arabic transliteration, in *Arabic Computational Morphology* (Springer, 2007), pp. 15–22

18. I. Guellil, A. Faical, Bilingual lexicon for Algerian Arabic dialect treatment in social media, in *WiNLP: Women & Underrepresented Minorities in Natural Language Processing (Co-located with ACL 2017)* (2017). http://www.winlp.org/wp-content/uploads/2017/final_papers_2017/92_Paper.pdf

19. M. Al-Badrashiny, R. Eskander, N. Habash, O. Rambow, Automatic transliteration of romanized dialectal Arabic, in *Proceedings of the Eighteenth Conference on Computational Natural Language Learning*, 2014, pp. 30–38

20. K. Meftouh, S. Harrat, S. Jamoussi, M. Abbas, K. Smaili, Machine translation experiments on PADIC: a parallel Arabic dialect corpus, in *The 29th Pacific Asia Conference on Language, Information and Computation*, 2015

21. G. Kumar, Y. Cao, R. Cotterell, C. Callison-Burch, D. Povey, S. Khudanpur, Translations of the Callhome Egyptian Arabic corpus for conversational speech translation, in *IWSLT*. Citeseer, 2014

22. R. Suwaileh, M. Kutlu, N. Fathima, T. Elsayed, M. Lease, Arabicweb16: a new crawl for today's Arabic web, in *Proceedings of the 39th International ACM SIGIR Conference on Research and Development in Information Retrieval* (ACM, 2016), pp. 673–676

23. M. Rushdi-Saleh, M.T. Martín-Valdivia, L.A. Ureña-López, J.M. Perea-Ortega, OCA: opinion corpus for Arabic. J. Assoc. Inf. Sci. Technol. **62**(10), 2045–2054 (2011)

24. N. Abdulla, S. Mohammed, M. Al-Ayyoub, M. Al-Kabi et al., Automatic lexicon construction for Arabic sentiment analysis, in *2014 International Conference on Future Internet of Things and Cloud (FiCloud)* (IEEE, 2014), pp. 547–552
25. M. Abdul-Mageed, M.T. Diab, AWATIF: a multi-genre corpus for modern standard Arabic subjectivity and sentiment analysis, in *LREC*. Citeseer, 2012, pp. 3907–3914
26. M. Aly, A. Atiya, LABR: a large scale Arabic book reviews dataset, in *Proceedings of the 51st Annual Meeting of the Association for Computational Linguistics (Volume 2: Short Papers)*, vol. 2 (2013), pp. 494–498
27. G. Badaro, R. Baly, H. Hajj, N. Habash, W. El-Hajj, A large scale Arabic sentiment lexicon for Arabic opinion mining, in *Proceedings of the EMNLP 2014 Workshop on Arabic Natural Language Processing (ANLP)*, 2014, pp. 165–173
28. S.R. El-Beltagy, NileULex: a phrase and word level sentiment lexicon for Egyptian and modern standard Arabic, in *LREC* (2016)
29. M. van der Wees, A. Bisazza, C. Monz, A simple but effective approach to improve Arabizi-to-English statistical machine translation, in *Proceedings of the 2nd Workshop on Noisy User-Generated Text (WNUT)*, 2016, pp. 43–50
30. J. May, Y. Benjira, A. Echihabi, An Arabizi-English social media statistical machine translation system, in *Proceedings of the 11th Conference of the Association for Machine Translation in the Americas*, 2014, pp. 329–341
31. I. Guellil, F. Azouaou, M. Abbas, Comparison between neural and statistical translation after transliteration of Algerian Arabic dialect, in *WiNLP: Women & Underrepresented Minorities in Natural Language Processing (Co-located with ACL 2017)*, 2017
32. I. Guellil, F. Azouaou, Neural vs statistical translation of Algerian Arabic dialect written with Arabizi and Arabic letter, in *The 31st Pacific Asia Conference on Language, Information and Computation PACLIC 31 (2017)*, 2017
33. R.M. Duwairi, M. Alfaqeh, M. Wardat, A. Alrabadi, Sentiment analysis for Arabizi text, in *2016 7th International Conference on Information and Communication Systems (ICICS)* (IEEE, 2016), pp. 127–132
34. I. Guellil, A. Adeel, F. Azouaou, A. Hussain, SentiALG: automated corpus annotation for Algerian sentiment analysis. arXiv preprint arXiv:1808.05079 (2018)
35. S. Medhaffar, F. Bougares, Y. Esteve, L. Hadrich-Belguith, Sentiment analysis of Tunisian dialects: linguistic resources and experiments, in *Proceedings of the Third Arabic Natural Language Processing Workshop*, 2017, pp. 55–61
36. I. Guellil, F. Azouaou, H. Saâdane, N. Semmar, Une approche fondée sur les lexiques d'analyse de sentiments du dialecte algérien (2017)
37. I. Guellil, F. Azouaou, F. Benali, A.-E. Hachani, H. Saadane, Approche hybride pour la translitération de l'arabizi algérien : une étude préliminaire, in *Conference: 25e conférence sur le Traitement Automatique des Langues Naturelles (TALN), May 2018, Rennes, FranceAt: Rennes, France* (2018). https://www.researchgate.net/publication/326354578_Approche_Hybride_pour_la_transliteration_de_l%27arabizi_algerien_une_etude_preliminaire
38. P. Koehn, H. Hoang, A. Birch, C. Callison-Burch, M. Federico, N. Bertoldi, B. Cowan, W. Shen, C. Moran, R. Zens et al., Moses: open source toolkit for statistical machine translation, in *Proceedings of the 45th Annual Meeting of the ACL on Interactive Poster and Demonstration Sessions* (Association for Computational Linguistics, 2007), pp. 177–180
39. S. Al-Azani, E.-S.M. El-Alfy, Using word embedding and ensemble learning for highly imbalanced data sentiment analysis in short Arabic text. Procedia Comput. Sci. **109**, 359–366 (2017)
40. A.A. Altowayan, L. Tao, Word embeddings for Arabic sentiment analysis, in *2016 IEEE International Conference on Big Data (Big Data)* (IEEE, 2016), pp. 3820–3825
41. A. El Mahdaouy, E. Gaussier, S.O. El Alaoui, Arabic text classification based on word and document embeddings, in *International Conference on Advanced Intelligent Systems and Informatics* (Springer, 2016), pp. 32–41
42. A. Barhoumi, Y.E.C. Aloulou, L.H. Belguith, *Document Embeddings for Arabic Sentiment Analysis* (2017)

43. A. Dahou, S. Xiong, J. Zhou, M.H. Haddoud, P. Duan, Word embeddings and convolutional neural network for Arabic sentiment classification, in *Proceedings of COLING 2016, the 26th International Conference on Computational Linguistics: Technical Papers*, 2016, pp. 2418–2427

44. M. Attia, Y. Samih, A. El-Kahky, L. Kallmeyer, Multilingual multi-class sentiment classification using convolutional neural networks, in *LREC* (2018)

45. R. Zbib, E. Malchiodi, J. Devlin, D. Stallard, S. Matsoukas, R. Schwartz, J. Makhoul, O.F. Zaidan, C. Callison-Burch, Machine translation of Arabic dialects, in *Proceedings of the 2012 conference of the North American Chapter of the Association for Computational Linguistics: Human Language Technologies* (Association for Computational Linguistics, 2012), pp. 49–59

46. W. Salloum, N. Habash, Dialectal to standard Arabic paraphrasing to improve Arabic-English statistical machine translation, in *Proceedings of the First Workshop on Algorithms and Resources for Modelling of Dialects and Language Varieties* (Association for Computational Linguistics, 2011), pp. 10–21

47. M. Taboada, J. Brooke, M. Tofiloski, K. Voll, M. Stede, Lexicon-based methods for sentiment analysis. Comput. Linguist. **37**(2), 267–307 (2011)

48. T. Mikolov, I. Sutskever, K. Chen, G.S. Corrado, J. Dean, Distributed representations of words and phrases and their compositionality, in *Advances in Neural Information Processing Systems*, 2013, pp. 3111–3119

49. Q. Le, T. Mikolov, Distributed representations of sentences and documents, in *International Conference on Machine Learning*, 2014, pp. 1188–1196

50. M. Abdul-Mageed, M.T. Diab, M. Korayem, Subjectivity and sentiment analysis of modern standard Arabic, in *Proceedings of the 49th Annual Meeting of the Association for Computational Linguistics: Human Language Technologies: Short Papers*, vol. 2 (Association for Computational Linguistics, 2011), pp. 587–591

51. K. Meftouh, N. Bouchemal, K. Smaïli, A study of a non-resourced language: the case of one of the Algerian dialects, in *The third International Workshop on Spoken Languages Technologies for Under-Resourced Languages-SLTU'12*, 2012

52. F.J. Och, H. Ney, A systematic comparison of various statistical alignment models. Comput. Linguist. **29**(1), 19–51 (2003)

53. K. Heafield, KenLM: faster and smaller language model queries, in *Proceedings of the Sixth Workshop on Statistical Machine Translation* (Association for Computational Linguistics, 2011), pp. 187–197

54. K. Papineni, S. Roukos, T. Ward, and W.-J. Zhu, BLEU: a method for automatic evaluation of machine translation, in *Proceedings of the 40th Annual Meeting on Association for Computational Linguistics Association for Computational Linguistics*, 2002, pp. 311–318

55. I. Guellil, F. Azouaou, Arabic dialect identification with an unsupervised learning (based on a lexicon). application case: Algerian dialect, in *2016 IEEE International Conference on Computational Science and Engineering (CSE) and IEEE International Conference on Embedded and Ubiquitous Computing (EUC) and 15th International Symposium on Distributed Computing and Applications for Business Engineering (DCABES)* (IEEE, 2016), pp. 724–731

Sentiment Analysis in Healthcare: A Brief Review

Laith Abualigah, Hamza Essam Alfar, Mohammad Shehab
and Alhareth Mohammed Abu Hussein

Abstract Sentiment analysis is one of data mining types that estimates the direction of personality's sentiment analysis within natural language processing. Analyzing the text computational linguistics are used to deduce and analyze mental knowledge of Web, social media and related references. The examined data quantifies the global society's attitudes or feelings via specific goods, people or thoughts and expose the contextual duality of the knowledge. Sentiment analysis used in different approaches such as products and services reviews. Also is used in healthcare, there is a huge volume of information about healthcare obtainable online, such as personal blogs, social media, and on the websites about medical issues rating that are not obtained methodically. Sentiment analysis provides many benefits such as using medical information to achieve the best result to increase healthcare quality. In this paper, sentiment analysis methods and techniques are presented that used in the medical domain.

Keywords Sentiment analysis · Data mining · Natural Language Processing (NLP) · Computational linguistics

1 Introduction

The online experience gives benefits to the people who are interested in such a product from the other people who wrote their sentiment analysis on such a product [1, 2]. There are many sources to find this huge amount of data online, such as social media sites, online forums, personal blogs, etc. including a wide scope of topics [3–6]. People discuss their healthcare cases on a lot of medical websites and forums, and they share their illness, indications, and drugs. The experience of medical centers that people visited, also shares the availability, services, pleasure, etc. [7, 8].

L. Abualigah (✉) · H. E. Alfar · M. Shehab · A. M. A. Hussein
Faculty of Computer Sciences and Informatics, Amman Arab University, Amman 11953, Jordan
e-mail: lx.89@yahoo.com

© Springer Nature Switzerland AG 2020
M. Abd Elaziz et al. (eds.), *Recent Advances in NLP: The Case of Arabic Language*, Studies in Computational Intelligence 874,
https://doi.org/10.1007/978-3-030-34614-0_7

It is very important to patients when they learn from other patients experience to make decisions about their medical issue. Such as choosing hospital, clinic, and medication. This information also benefits the hospitals to know the patients' interests and problems and resolve them. Patients share their experience covered in their own sentiment analysis and passions, which is the power of this type of analysis. [9] has taught sentiment analysis as knowing the sentiments of people about a subject and its features. The medical content that is accessible online is free, in addition to its existing in large volume; therefore, analyzing this huge amount of data manually is less effective.

Assessment examination for the most part centered on the programmed acknowledgment of suppositions' extremity, as positive or negative. These days, notion investigation is supplanting the online and customary overview techniques normally led by organizations for finding the general conclusion about their items and administrations to improve their promoting methodology and item notice and help to improve user administration. The online accessibility of enormous content makes it imperative to be broke down. The programmed examination of this data includes a deep comprehension of common dialects. Notions and feelings assume a significant job in our day by day lives. They aid basic leadership, learning, correspondence, and circumstance mindfulness in human situations. The significance of handling and understanding vernacular content is expanding because of the development of socially produced regional substance in web-based life. Notwithstanding existing materials, for example, nearby sayings, exhortation, and fables that are discovered spread on the web [10].

Techniques in sentiment analysis complete this task with automatic processes without people help. In the past, researchers were used questionnaires and surveys for sentiment analysis purposes, and these methods taking a lot of money and time. There are a few papers produced by specialists, and they do not solve the problems, which the patient suffers. Sentiment analysis takes interests the feelings of patients stated in a huge number of documents that is shared on different websites. The main techniques used in sentiment analysis are shown in Fig. 1.

The result of sentiment analysis is the categorization of medical decisions into two types such as good, or not good. However, we can derive characters of medical problems by digging deeper. The purpose of this article is to focus on the importance of sentiment analysis represented by a huge number of patients regarding their disease, medications, medical issues, etc. As well, we added comprehensive details in this topic for who are interested in this domain in futures with various directions of research.

The rest of this paper is summarized as some of the related work in Sect. 3, and we talk about future work in Sect. 4, in Sect. 5 the related works are discussed and overviewed the used techniques, in Sect. 6 we will discuss some natural language processing challenges, and finally the conclusion and references in Sect. 6.

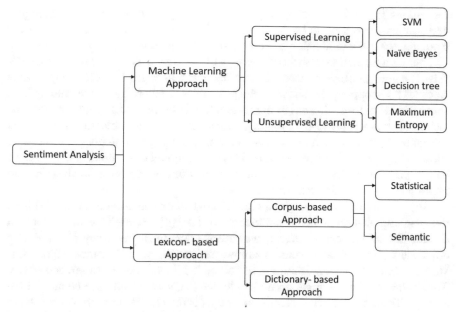

Fig. 1 The main techniques used in sentiment analysis

2 Related Work

In this section, we cover some important parts in the sentiment analysis as follows:

2.1 Sentiment Analysis and Sentiment Analysis Mining

An essential part of the information-gathering form has eternally been to find out what different people believe. With the increasing availability and demand of sentiment analysis-rich devices such as online survey sites and private blogs, unique events and difficulties occur as people now can, and do, actively use learning technologies to attempt out and get the sentiment analysis of others. The unexpected eruption of movement in the field of sentiment analysis mining and sentiment analysis, which dispenses with the computational method of sentiment analysis, feeling, and subjectivity in the document, has thus happened at least in part as a right answer to the surge of investment in new systems that distribute direct with ideas as a first-class object [11].

Web is the most critical wellspring of getting up considerations, overviews for an item, and surveys for an administration or movement. A Bulky measure of surveys are created day by day on the internet about online items and articles. For instance, numerous people share their comments, surveys, and emotions in their language using online life systems, for example, twitter, etc. The prerequisite for structuring systems for various lingos is extending, especially as blogging and small-scale blogging destinations are getting to be prominent [12]. Exhibited a complete review of the methodologies of examination and diagrams of significant gaps in the writing.

There is a method to analyze the feelings of humans, which is the sentiment analysis. The internet provides accessible textual data that is growing every day. Online websites for shopping give shoppers the possibility to record their notes and reviews about goods sold, and that has to improve products selling and to enhance the satisfaction of shoppers [8, 13–15]. Dealing with a huge amount of records or comments in sentiment analysis is hard because the process using this amount is difficult to extract the general sentiment analysis. This huge amount of sentiment analysis can't be analyzed manually, hence the automatic method of sentiment analysis has an important role in solving this problem.

[16] proposed another weighing plan for content examination purposes, [2] used the weighing plan for content examination. Just as, [4] proposed awaiting approach combined with the term frequency and inverse document frequency (TF*IDF) for Arabic assessment arrangement on administrations' audits in Lebanon nation. Surveys are about open administrations, including lodgings, eateries, shops, and others. They collected the materials from Google and Zomato, which have come to 3916 surveys. Trials show three center discoveries: (1) the classifier is sure when used to anticipate positive audits. (2) The model is one-sided on anticipating audits with a negative feeling. Finally, the low level of negative audits in the corpus adds to the modesty of the calculated relapse model. Other optimization techniques can be sued to solve this problem [17–25].

2.2   Sentiment Analysis for Health Care

In this work [26], the aspects of sentiment in the medical field and possible use-cases are reviewed. Through the research review, the state of the art in healthcare environments is summarized. To learn the linguistic peculiarities of thought in medical texts and to get open research issues of sentiment analysis in medicine, they make a quantitative evaluation concerning term language and sentiment combination of a dataset of clinical stories and medical group media obtained from six separate references.

Sentiment analysis in healthcare deals with the healthcare problems of patients [7]. The sentiment analysis of patients is taken to solve their problems quickly and this helps the decision-makers to make plans and useful changes. Sentiment analysis is used in different fields. Healthcare analysis-based study displays the powerful points of medicines and services [1, 5].

2.3 Sentiment Lexicons for Health-Related Sentiment Analysis Mining

Natural language processing includes an approach called sentiment analysis mining; this approach recognizes the passionate tone behind a frame of the document. This is a popular way for organizations to determine and categorize sentiment analysis about a product, service or idea [27–34].

For this task, there are some resources required like polarized lexicon. Sentiment analysis mining in healthcare is not explored well, partially because some confidence is provided to patients and their sentiment analysis, and there are many patients used social media. They are inspired in sentiment analysis mining of reviews about medications. Firstly, they define the origin of lexicon, including sentiment analysis words from the common domain and their polarity [35]. Then they perform the production of a medical sentiment analysis lexicon, based on a corpus of medication reviews.

The Arabic language experiences the absence of accessible enormous datasets for AI and sentiment analysis applications. This work introduced a huge dataset called BRAD, which is the biggest book reviews in Arabic Dataset [36]. This dataset contains 490,587 inn surveys gathered from the Booking.com site. Each record contains the survey message in the Arabic language, the commentator's evaluating on a size of 1 to 10 stars, and different qualities about the lodging/analyst. They make accessible the full unequal dataset just as a decent subset. To analyze the datasets, six prevalent classifiers are utilized using Modern Standard Arabic (MSA). They test the slant analyzers for extremity and rating groupings. Moreover, we actualize an extremity dictionary-based slant analyzer. The discoveries affirm the viability of the classifiers and the datasets. Our center commitment is to make this benchmark-dataset accessible and available to the examination network on the Arabic language.

Information explanation is important to sort content into two classes, the first thing, a decent comment rule is needed, and the second is building up what is required to fit the bill for each class. In this paper [37], they present a novel way to deal with naturally developed corpus for Algerian tongue (A Maghrebi Arabic vernacular). The development of this corpus depends on an Algerian feeling vocabulary that is likewise built naturally. They worked on two broadly utilized contents using Arabic internet-based life: Arabic and Arabizi. The accomplished F1-score is up to 72% and 78% for an Arabic and Arabizi test, respectively.

2.4 Sentiment Analysis on Health Services

Social media system appearance produces huge numbers of important data that is accessible online and accessible easily. Several users share discussions, photos, and videos, data, and viewpoints on various social media websites, within Twitter presence one of the common modern [9, 38, 39].

Consistently a lot of emotional data is created through informal organizations, for example, Facebook and Twitter. The abstract data suggests the suppositions, convictions, emotions, and frames of mind that individuals express towards various subjects of intrigue [40]. In addition, this sort of data is critical for organizations, associations or people, since it enables them to do activities that advantage them. In addition, assessment examination is the field that reviews emotional data through characteristic language handling, computational etymology, data recovery, and information mining strategies. Sentiment analysis is exceptionally helpful in different spaces, for example, legislative issues, highlighting, the travel industry, among others. In fact, the social insurance space suggests a huge region of the chance to acquire advantages utilizing sentiment analysis examination, for example, getting data about the patients' state of mind, illnesses, unfriendly medication responses, plagues, among others. Nevertheless, the social insurance space has been almost no investigated.

It is very difficult to extract useful information from Twitter, because this data is unregulated, and this a big challenging responsibility. There are many Arab users on twitter, and these users are used the Arabic language to write their posts and tweets. In the English language, there are a lot of researches about sentiment analysis, but in the Arabic language, it is very poor. They introduced an Arabic language dataset about sentiment analysis on medical services received from Twitter [41].

2.5 Sentiment Analysis Techniques in Healthcare

A large number of subjective knowledges is created through social media websites such as Facebook, Twitter. This knowledge indicates the ideas, views, emotions, and beliefs that people pass across various issues of matter. It has a revolutionary point for several organizations or people in order to provide information about what represented to the goods or services, which allows them to carry out activities that make advantages for them, such as how to make better choices, better promotion operations or business strategies, among others [42].

Individuals are uncommon to discuss about their medical issues with one another and, it is extremely poor to see about their practical wellbeing circumstance. Concentrate just on Twitter, clients' made tweets made out of the news, governmental issues, life discussion which can likewise be connected for completing an assortment of examination purposes. Hence, the human services framework is created to help the experts to effectively check their conduct sentiment analysis depend on Twitter information. Most extreme Entropy classifier (MaxEnt) is utilized to perform sentiment analysis on their tweets to propose their wellbeing condition (great, reasonable, or terrible). It is communicating with Twitter information (huge information condition) thus, the Internet of Things (IoT) based huge information preparing structure is worked to be effectively dealt with a lot of Twitter client' information [43]. The method of personal knowledge and the knowledge to recognize the feelings and passions expressed in social networks needs sentiment analysis. They defined sentiment analysis as the consideration of the feelings, sentiments, and emotions that people come across various points of concern. This investigation utilizes natural language processing and uses text analysis, in addition to computational linguistics, and it includes duties of exposure, descent, and division of sentiments on separate devices like treatment discussions on the web, and social networks.

2.6 Sentiment Analysis for Arabic Language

Arabic sentiment analysis being very interesting domain. Generally, sentiment analysis has several papers in English, the Arabic language is still in its early levels in this area. In this paper [44], they investigated an Arabic sentiment analysis application by performing a sentiment analysis for Arabic tweets. The obtained tweets are investigated to produce their feelings polarity (positive, or negative).

A new method is proposed in [35] for subjectivity and sentiment analysis in Arabic social media language. The Arabic language is a morphologically strong language, which gives important complexities for conventional approaches to making the proposed method designed for the English language. Apart from the challenges offered by the social media genres processing, the Arabic language naturally has a high amount of shifting word forms beginning to data sparsity. Albeit informal organizations have turned into an important asset for mining sentiment analysis, there is no past research exploring the nonprofessional's feeling towards Spanish expressions of Arabic historical underpinnings identified with Islamic phrasing. [37] aimed at analyzing Spanish expressions of Arabic beginning identified with Islam. An irregular example of 4586 out of 45,860 tweets was utilized to assess the general notion towards some Spanish expressions of Arabic starting point identified with Islam. A specialist predefined Spanish vocabulary of around 6800 seed descriptors was utilized to direct the examination. Results demonstrate a for the most part positive supposition towards a few Spanish expressions of Arabic derivation identified with Islam. By executing both a subjective and quantitative strategy to investigate tweets' estimations towards Spanish expressions of Arabic derivation, this examination adds broadness and profundity to the discussion over Arabic phonetic effect on Spanish vocabulary.

In this meaning, they address the next 4 pertinent problems: how to best serve lexical information; whether standard characteristics used for English are helpful for Arabic; how to handle Arabic languages; and, whether genre-specific characteristics have a calculable impression on performance. The outcomes show that using either lemma or lexeme data is important, as well as utilizing the two-part of speech tagsets (RTS and ERTS). Nonetheless, the outcomes present that they need individualized resolutions for each genre and job, but that lemmatization and the ERTS POS tagset are already in a bulk of the frames [43]. Several studies have been used sentiment analysis in Arabic such as in [45].

The study [13] tested four unique dictionaries: an interpretation of Harvard IV-4 Dictionary (Harvard), interpretation of the MPQA subjectivity vocabulary created by Pittsburgh University (HRMA) and two distinct executions of MPQA. We assessed every one of the four vocabularies with three datasets from various spaces, one of them is about wellbeing remarks (PatientJo), the second is from Twitter information, and the third is about books surveys (LABR). For sentence-level, they proposed six unique techniques for assumption esteems and report extremity. The outcomes demonstrate that the HRMA vocabulary performs superior to different dictionaries in LABR while Harvard performs better in the patient dataset. The outcomes demonstrate that the dictionary-based methodology for record level and sentence-level techniques produce a comparative execution. They saw that giving additional load for the first and last sentences in sentence-level methodology improves the general execution as far as exactness.

Under neural networks, good data skills are given during dealing with complex and huge datasets from a wide variety of utilization fields. Deep Convolutional Neural Networks (CNNs) give benefits in choosing useful features and Long Short-Term Memory (LSTM) networks demonstrated excellent capabilities to get constant data. Also, the current pre-processing mechanisms for the Arabic language is another weakness, alongside with insufficient research prepared in this field. The privileges of combining CNNs and LSTMs are analyzed in [46] and the detonation achieved enhanced precision for Arabic sentiment analysis on various datasets. Moreover, they attempt to analyze the morphological difference of special Arabic terms by applying various sentiment classification approaches.

The difficulties rest in the fact that most Arabic users compose unorganized idiom texts rather than writing in common conventional Arabic. In [47], these difficulties by matching among two approaches: utilizing sentiment analysis methods directly on the language; and utilizing a key that transforms from accent to the common conventional Arabic text, then producing a sentiment analysis on the outcomes common conventional Arabic text. Eventually, Saudi Twitter data is examined in this paper.

Applying Convolutional Neural Networks (CNNs) for Sentiment Analysis (SA) has got better results. CNN's are groundbreaking at separating a various leveled portrayal of the contribution by stacking numerous convolutional and pooling layers. In this paper [16], two deep CNNs are connected for Arabic supposition examination utilizing character-level highlights as it were. An enormous scale dataset is built from accessible SA datasets to prepare systems. The dataset keeps up feelings from various spaces communicated in various Arabic structures (Modern Standard, Dialectal). Other than various AI calculations as Logistic Regression, Support Vector Machine and Naïve Bayes have been connected to survey the exhibition on such a huge dataset. Up to the accessible learning, this is the primary utilization of character level deep CNNs for Arabic language supposition examination. Results demonstrate the capacity of Deep CNNs models to arrange Arabic feelings relying upon character portrayal just and register 7% improved precision contrasted with AI classifiers.

3   Future Work

Sentiment analysis methods would produce aggregated personal decisions on the healthcare inquiries. A suggested way can be made on top of such inquiry analysis that would suggest medications, procedures, experts in the region, important health care stations etc. based on the individual information presented by the different patients.

More complex methods can be used in such methods to control spam. The use of science, fundamental knowledge and machine learning methods must merge to decrease the influence on decisions. More such parameters are to be determined, through which the efficacy of content can be checked. Producers may also be assigned based on the legality of the content they provide where producers with technical experience may be overlooked or viewed more accurately for later contents.

4   Challenges of the Natural Language Processing

The Natural Language Processing has several difficulties that can change the appearance of the sentiment analysis in many aspects [48]. Some of the certain difficulties are related to the kind of data while others are obvious to any type of analyzing text. The current difficulties in Natural Language Processing can be divided as follows [35]:

The document level difficulties are associated with the inquiry text that can have reviews, which are only found in blogs. Blog reports provide annotations that normally become within a forum. These reports have feelings that are particular to the field. Based on personal therapy and diversity in natural languages, personals display themselves negatively. Sentiment analysis spamming is also a very sensitive point where somebody gives incorrect sentiment analysis prepared for serving or reducing special-purpose things (manually). There may also be displays as inspection reports that would become nothing to do with the destination actuality in the discussion [49].

Regrettably, there is somebody and organizations included in the market of sentiment analysis study spamming. Specialists in sentiment analysis are demand as diverse as half of the inspections to be spammed on any popular display roots. Utilization of special sentiment indications are also required as the authors' content. Finally, grammar errors, local slangs are other generally handled difficulties.

5   Discussions and Overview

There are many challenges facing sentiment analysis techniques. For instance, the complexity in the way of the people to express sentiment analysis, lexical content in the text, irony, and implication. Therefore, various techniques have been used

Table 1 Shows the various techniques have been used in sentiment analysis

Author(s)	Technique	Approach	Accuracy
[49]	Lexicon	Corpus- based Approach	54%
[43]	SVM Naïve Bayes Maximum Entropy	Supervised	
[50]	Max Entropy	Supervised	Positive (34.34%) Negative (22.18%) Neutral (43.46%)
[16]	SVM Naïve Bayes	Supervised	94.18%
[50]	Logistic Regression	Supervised	
[12]	SVM Decision tree Nearest Neighbor (k-NN)	Supervised	Below 50%
[1]	Lexicon	Corpus- based Approach	
[13]	Lexicon	Corpus- based Approach	
[37]	Lexicon	Corpus- based Approach	78%
[40]	SVM Decision tree	Supervised	
[46]	Convolutional Neural Networks (CNNs)	Supervised	91%

(see Table 1) to solve the sentiment analysis problems. However, each technique has advantages and disadvantages. For example, the Naïve Bayes is efficient and fast computation without influenced by irrelevant features. However, it assumes independent attributes. While Maximum Entropy does not suppose statistical independence of random elements [50]. However, it needs more efforts from the human in the form of additional resource [7]. Finally, the main advantage of the lexicon is that since the accuracy of the comments achieved by humans is not guarantee. Nevertheless, this technique consumes immense time [1].

6 Conclusion

Sentiment analysis is a necessary way to help people in getting a recommendation and read knowledge. This technique aims to analyze the social media, wherever a problem highlighted may only contact the necessary authorities if they notice it immediately. It is impossible through the social media and various user content to get the right recommendations. Sentiment analysis is automating this process. Sentiment analysis aim to get more information to assist users to get the right decision about the studied. The sentiment analysis methods are applied to data mining and machine learning to adjust this difficulty. Supervised procedures with high precision can be employed to further sensible proposals for finding close bound problems. Unsupervised techniques are less costly and can be used to investigate big data. The performance of the sentiment analysis process is categorized with interest rates for the potential sources. It can be supported with graphical devices to be more decisive to users. Reviews may additionally be applied to set feather the conclusions. This state is still greatly from being ready with new sub-streams known as disturbance analysis, expression analysis, preference analysis, risk analysis, etc. The demand and insufficiency of these decisions can be recognized from the evidence that it has previously been done for financial goods while it is still active as a research difficulty. Finally, the optimization way can be used to deal with this problem; it gave promising results in solving several problems.

References

1. M.T. Khan, S. Khalid, Sentiment analysis for health care, in *Big Data: Concepts, Methodologies, Tools, and Applications*, IGI Global (2016), pp. 676–689
2. M. Shehab, A.T. Khader, M.A. Al-Betar, L.M. Abualigah, Hybridizing cuckoo search algorithm with hill climbing for numerical optimization problems, in *2017 8th International Conference on Information Technology (ICIT)*, IEEE (2017, May), pp. 36–43
3. L.M.Q. Abualigah, Feature selection and enhanced Krill Herd algorithm for text document clustering, in *Studies in Computational Intelligence* (2019)
4. H. Mulki, H. Haddad, C. Bechikh Ali, I. Babaoğlu, Tunisian dialect sentiment analysis: a natural language processing-based approach. Computación y Sistemas **22**(4) (2018)
5. M.M. Mostafa, N.R. Nebot, Sentiment analysis of spanish words of arabic origin related to islam: a social network analysis. J. Lang. Teach. Res. **8**(6), 1041–1049 (2017)
6. G. Vinodhini, R.M. Chandrasekaran, Sentiment analysis and sentiment analysis mining: a survey. Int. J. **2**(6), 282–292 (2012)
7. H. Iyer, M. Gandhi, S. Nair, Sentiment analysis for visuals using natural language processing. Int. J. Comput. Appl. **128**(6), 31–35 (2015)
8. S.N. Manke, N. Shivale, A review on: sentiment analysis mining and sentiment analysis based on natural language processing. Int. J. Comput. Appl. **109**(4) (2015)
9. F. Greaves, D. Ramirez-Cano, C. Millett, A. Darzi, L. Donaldson, Use of sentiment analysis for capturing patient experience from free-text comments posted online. J. Med. Internet Res. **15**(11), e239 (2013)
10. I.O. Hussien, K. Dashtipour, A. Hussain, Comparison of sentiment analysis approaches using modern Arabic and Sudanese Dialect, in *International Conference on Brain Inspired Cognitive Systems* (Springer, Cham, 2018, July), pp. 615–624
11. B. Pang, L. Lee, Sentiment analysis mining and sentiment analysis. Found. Trends® Inf. Retrieval **2**(1–2), 1–135 (2008)
12. N. Mukhtar, M.A. Khan, Urdu sentiment analysis using supervised machine learning approach. Int. J. Pattern Recognit. Artif. Intell. **32**(02), 1851001 (2018)
13. H. Awwad, A. Alpkocak, Performance comparison of different lexicons for sentiment analysis in Arabic, in *2016 Third European Network Intelligence Conference (ENIC)*, IEEE (2016, September), pp. 127–133
14. B. Liu, Sentiment analysis and sentiment analysis mining. Synth. Lect. Human Lang. Technol. **5**(1), 1–167 (2012)
15. A. Pak, P. Paroubek, Twitter as a corpus for sentiment analysis and sentiment analysis mining. LREc **10**, 1320–1326 (2010)

16. M. Shehab, A.T. Khader, M.A. Alia, Enhancing Cuckoo search algorithm by using reinforcement learning for constrained engineering optimization problems, in *2019 IEEE Jordan International Joint Conference on Electrical Engineering and Information Technology (JEEIT)*, IEEE (2019, April), pp. 812–816

17. L.M. Abualigah, A.T. Khader, E.S. Hanandeh, A new feature selection method to improve the document clustering using particle swarm optimization algorithm. J. Comput. Sci. (2017)

18. L.M. Abualigah, A.T. Khader, E.S. Hanandeh, A combination of objective functions and hybrid Krill Herd algorithm for text document clustering analysis, in *Engineering Applications of Artificial Intelligence* (2018)

19. L.M. Abualigah, A.T. Khader, E.S. Hanandeh, A novel weighting scheme applied to improve the text document clustering techniques, in *Innovative Computing, Optimization and Its Applications* (Springer, Cham, 2018), pp. 305–320

20. L.M. Abualigah, A.T. Khader, E.S. Hanandeh, Hybrid clustering analysis using improved krill herd algorithm. Appl. Intell. (2018)

21. L.M. Abualigah, A.T. Khader, M.A. Al-Betar, O.A. Alomari, Text feature selection with a robust weight scheme and dynamic dimension reduction to text document clustering. Expert Syst. Appl. **84**, 24–36 (2017)

22. L.M. Abualigah, A.T. Khader, M.A. AlBetar, E.S. Hanandeh, Unsupervised text feature selection technique based on particle swarm optimization algorithm for improving the text clustering, in *Eai International Conference on Computer Science and Engineering* (2017)

23. L.M. Abualigah, A.T. Khader, E.S. Hanandeh, A.H. Gandomi, A novel hybridization strategy for krill herd algorithm applied to clustering techniques. Appl. Soft Comput. **60**, 423–435 (2017)

24. K. Denecke, Y. Deng, Sentiment analysis in medical settings: new opportunities and challenges. Artif. Intell. Med. **64**(1), 17–27 (2015)

25. Z.A. Al-Sai, L.M. Abualigah, Big data and E-government: a review, in *2017 8th International Conference on Information Technology (ICIT)*, IEEE (2017, May), pp. 580–587

26. M.Z. Asghar, A. Khan, F.M. Kundi, M. Qasim, F. Khan, R. Ullah, I.U. Nawaz, Medical sentiment analysis lexicon: an incremental model for mining health reviews. Int. J. Acad. Res. **6**(1), 295–302 (2014)

27. M.Z. Asghar, M. Qasim, B. Ahmad, S. Ahmad, A. Khan, I.A. Khan, Health miner: sentiment analysis extraction from user generated health reviews. Int. J. Acad. Res. **5**(6), 279–284 (2013)

28. S. Baccianella, A. Esuli, F. Sebastiani, Sentiwordnet 3.0: an enhanced lexical resource for sentiment analysis and sentiment analysis mining. Lrec **10**, 2200–2204 (2010)

29. E. Cambria, B. Schuller, Y. Xia, C. Havasi, New avenues in sentiment analysis mining and sentiment analysis. IEEE Intell. Syst. **28**(2), 15–21 (2013)

30. A. Carvalho, A. Levitt, S. Levitt, E. Khaddam, J. Benamati, Off-the-shelf artificial intelligence technologies for sentiment and emotion analysis: a tutorial on using IBM natural language processing. Commun. Assoc. Inf. Syst. **44**(1), 43 (2019)

31. M. Chaudhari, S. Govilkar, A survey of machine learning techniques for sentiment classification. Int. J. Comput. Sci. Appl. **5**(3), 13–23 (2015)

32. N. Godbole, M. Srinivasaiah, S. Skiena, Large-scale sentiment analysis for news and blogs. Icwsm **7**(21), 219–222 (2007)

33. M.D. Hauser, N. Chomsky, W.T. Fitch, The faculty of language: what is it, who has it, and how did it evolve. Science **298**(5598), 1569–1579 (2002)

34. A. Elnagar, Y.S. Khalifa, A. Einea, Hotel Arabic-reviews dataset construction for sentiment analysis applications, in *Intelligent Natural Language Processing: Trends and Applications* (Springer, Cham, 2018), pp. 35–52

35. I. Guellil, A. Adeel, F. Azouaou, A. Hussain, Sentialg: automated corpus annotation for algerian sentiment analysis, in *International Conference on Brain Inspired Cognitive Systems* (Springer, Cham, 2018, July), pp. 557–567

36. A.M. Alayba, V. Palade, M. England, R. Iqbal Improving sentiment analysis in Arabic using word representation, in *2018 IEEE 2nd International Workshop on Arabic and Derived Script Analysis and Recognition (ASAR)*, IEEE (2018, March), pp. 13–18

37. P. Gonçalves, M. Araújo, F. Benevenuto, M. Cha, Comparing and combining sentiment analysis methods, in Proceedings of the First ACM Conference on Online Social Networks (ACM, 2013, October), pp. 27–38
38. F.J. Ramírez-Tinoco, G. Alor-Hernández, J.L. Sánchez-Cervantes, M. del Pilar Salas-Zárate, R. Valencia-García, Use of sentiment analysis techniques in healthcare domain, in *Current Trends in Semantic Web Technologies: Theory and Practice* (Springer, Cham, 2019), pp. 189–212
39. E. Refaee, V. Rieser An arabic twitter corpus for subjectivity and sentiment analysis, in *LREC* (2014, May), pp. 2268–2273
40. M. Al-Ayyoub, A.A. Khamaiseh, Y. Jararweh, M.N. Al-Kabi, A comprehensive survey of arabic sentiment analysis. Inf. Process. Manage. **56**(2), 320–342 (2019)
41. H. Htet, S.S. Khaing, Y.Y. Myint, Tweets sentiment analysis for healthcare on big data processing and IoT architecture using maximum entropy classifier, in *International Conference on Big Data Analysis and Deep Learning Applications* (Springer, Singapore, 2018, May), pp. 28–38
42. A. Shoukry, A. Rafea, Sentence-level Arabic sentiment analysis, in *2012 International Conference on Collaboration Technologies and Systems (CTS)*, IEEE (2012, May), pp. 546–550
43. M. Korayem, D. Crandall, M. Abdul-Mageed, Subjectivity and sentiment analysis of arabic: a survey, in *International Conference on Advanced Machine Learning Technologies and Applications* (Springer, Heidelberg, 2012, December), pp. 128–139
44. A.M. Alayba, V. Palade, M. England, R. Iqbal, A combined CNN and LSTM model for arabic sentiment analysis, in *International Cross-Domain Conference for Machine Learning and Knowledge Extraction* (Springer, Cham, 2018, August), pp. 179–191
45. S. Rizkallah, A. Atiya, H.E. Mahgoub, M. Heragy, Dialect versus MSA sentiment analysis, in *International Conference on Advanced Machine Learning Technologies and Applications* (Springer, Cham, 2018, February), pp. 605–613
46. A. Rajput, *Natural Language Processing, Sentiment Analysis and Clinical Analytics*. arXiv preprint arXiv:1902.00679 (2019)
47. A.N. Langville, C.D. Meyer, *Google's PageRank and Beyond: The Science of Search Engine Rankings*. Princeton University Press (2011)
48. L. Igual, S. Seguí, Statistical natural language processing for sentiment analysis, in *Introduction to Data Science* (Springer, Cham, 2017), pp. 181–197
49. L. Goeuriot, J.C. Na, W.Y. Min Kyaing, C. Khoo, Y.K. Chang, Y.L. Theng, J.J. Kim, Sentiment lexicons for health-related sentiment analysis mining, in *Proceedings of the 2nd ACM SIGHIT International Health Informatics Symposium* (ACM, 2012, January), pp. 219–226
50. M. Al Omari, M. Al-Hajj, N. Hammami, A. Sabra, Sentiment classifier: logistic regression for Arabic services' reviews in Lebanon, in *2019 International Conference on Computer and Information Sciences (ICCIS)*, IEEE (2019, April), pp. 1–5

Aspect-Based Sentiment Analysis for Arabic Government Reviews

Sufyan Areed, Omar Alqaryouti, Bilal Siyam and Khaled Shaalan

Abstract Government services are available online and can be provided through multiple digital channels, clients' feedback on these services can be submitted and obtained online. Enormous budgets are invested annually by governments to understand their clients and adapt services to meet their needs. In this paper, a unique dataset that consists of government smart apps Arabic reviews, domain aspects and opinion words is produced. It illustrates the approach that was carried out to manually annotate the reviews, measure the sentiment scores to opinion words and build the desired lexicons. Furthermore, this paper presents an Arabic Aspect-Based Sentiment Analysis (ABSA) that combines lexicon with rule-based models. The proposed model aims to extract aspects of smart government applications Arabic reviews, and classify all corresponding sentiments. This model examines mobile government app reviews from various perspectives to provide an insight into the needs and expectations of clients. In addition, it aims to develop techniques, rules and lexicons for language processing to address variety of SA challenge. The performance of the proposed approach confirmed that applying rules settings that can handle some challenges in ABSA improves the performance significantly. The results reported in the study have shown an increase in the accuracy and f-measure by 6%, and 17% respectively when compared with the baseline.

Keywords Aspect-Based Sentiment Analysis (ABSA) · Government smart apps reviews · Lexicon-based · Rule-based · Arabic reviews · Dubai Government

S. Areed (✉) · O. Alqaryouti · B. Siyam
The British University in Dubai, Dubai, UAE
e-mail: sufyanareed@gmail.com

O. Alqaryouti
e-mail: omar.alqaryouti@gmail.com

B. Siyam
e-mail: siyam.bilal@gmail.com

K. Shaalan
School of Informatics, University of Edinburgh, Edinburgh, UK
e-mail: khaled.shaalan@buid.ac.ae

© Springer Nature Switzerland AG 2020
M. Abd Elaziz et al. (eds.), *Recent Advances in NLP: The Case of Arabic Language*, Studies in Computational Intelligence 874,
https://doi.org/10.1007/978-3-030-34614-0_8

143

1 Introduction

The dramatic improvement in the internet evolution, and the spread of smart mobile devices in addition to the uncountable number of mobile applications, have caused a massive increase in the amount of unstructured text. Therefore, there was a growing need to find an analytic process that enables analysts to restructure text which will result in understanding humans' opinions. Humans are subjective creatures, and can express their feelings in a complicated manner. On the other hand, their opinions will influence any decision they may take. To be able to understand these opinions from a tremendous amount of text, it becomes necessary to make information systems recognize people on that level, which requires additional efforts to be placed by researchers.

One of the classification levels in Sentiment Analysis is aspect-based. It consists of feature or aspect extraction, polarity predication of sentiment (classification), and aggregation of sentiment [1, 2]. Aspect extraction includes the review aspects recognition that can be identified through the clients' comments in order to extract the aspects or features that are desired. It is also necessary to specify the entity to which the aspect belongs [3]. For instance, a review of "Dubai Police" mobile app: "Dubai Police application is well designed", where "design" can be considered as an aspect of "Dubai Police" entity. The sentiment therefore can reflect this particular aspect. Then, a prediction and classification of polarity are carried out to determine whether the aspect sentiment polarity indicates a positive, negative or neutral orientation in addition to the strength level. In the given example, the word "well" indicates a positive sentiment polarity for the "design" aspect. Finally, the results are summarized in line with the extracted aspects and their corresponding classified polarity.

[4] Showed a mockup that allows the user to equate and envision the advantages and disadvantages of products' aspects. On the other hand, an aggregated opinions summary of a traditional text-based was produced by [5] which could provide a brief overview of the people's judgment about a product. The disadvantage of this approach is that it provides a qualitative summary only, in comparison to quantitative summaries which can give users a simple, analytical, graphical and concise options. Classification of SA aspects can be explicit and implicit. The aspects terms of subjective sentences clearly reflect the opinion objective in explicit aspects, while the aspects terms in implicit aspects do not reflect the opinion objective clearly but will be extracted through the semantic of the sentence [3, 4, 6–8]. The previous example is showing an explicit aspect "design" with a positive sentiment polarity "well". In contrast, another review e.g. "The app operates according to expectations" which cannot represent any explicit term that indicate explicit aspect. Instead, "according to expectations" implicitly indicates a positive sentiment polarity that is related to "Functionality and Performance".

Aspect extraction is a primary task in SA aimed at extracting sentimental objectives from the context of opinion [3]. Governments spend enormous budgets to understand public and listen to their feedback voice, the provision of e-services that satisfy the needs and expectations of citizens is vital to have a proper e-government bodies. This can be achieved by exploring important sources of data which can help to have a comprehensive framework [9]. The extraction of aspects in SA has become an energetic research topic, since it is the most important task to recognize aspects [8, 10].

In this paper, a unique dataset that consists of government smart apps Arabic reviews, domain aspects and opinion words is produced. It illustrates the approach that was carried out to manually annotate the reviews, measure the sentiment scores to opinion words and build the desired lexicons. Moreover, this study aims to construct an integrated lexicon-based with a model that is driven from rule-based. The model objective is to extract explicit and implicit aspects and classify the Arabic smart government apps review into the corresponding sentiments. The approach proposed can contribute in highlighting multiple challenges that are related to unstructured texts with noise compared to other approaches. The proposed solution can help government entities acquire a holistic view of their customers' needs and level of expectations. In addition, the model offers government agencies insights into the performance of their smart apps.

There are 7 sections in this paper. In Sect. 2, the related works are discussed in the literature with regards to approaches related to dataset contraction and aspect extraction from unstructured text resources. Next, Sect. 3 describes our approach in constructing the dataset. Section 4 describes our approach in extracting aspects and classifying sentiments. Section 5 discusses the results and performance comparison. Section 6 discusses the theoretical and practical implications of our work. Finally, in Sect. 7 the conclusion of our work is presented.

2 Related Work

There are different ways to build a Sentiment Lexicon. Some lexicons have been constructed manually based on dictionaries and annotated by a linguistic experts or native speakers, while some other lexicon have been constructed automatically through an association, where the score for each new word seeds calculated via considering the frequency of its proximity to preexisting seed words, were others built based on a semi-automatically constructed lexicon which requires manual intervention to normalize an automatically-built lexicon [11].

[12] have developed a corpus for movie reviews, the corpus consists of 500 movie reviews that have been annotated using support vector machine (SVM) and Naïve Bayes (NB) algorithms to set the opinion polarity of the reviews. The polarity ratio is balanced i.e. 250 positive reviews and 250 negative reviews, but data set was domain-specific and limited in size.

[13] Constructed an MSA corpus based on annotating 2855 sentences manually from the 1st 400 documents of Part 1 V 3.0 of the PATB. Annotation was done based on subjectivity and domain at sentence level. It was classified into four categories: objective, subjective-positive, subjective-negative and subjective-neutral.

[14] Presented AWATIF, a multi-genre corpus (documents from the newswire, Wikipedia Talk Pages, and Web Forums), where sentences have been annotated manually by three groups of linguistics students who attended a training for clear guideline, and crowdsourcing. The purpose of making 3 groups was to consider and evaluate the genre nuances. The analysis was conducted at sentence level as positive, neutral and negative.

[15] came up with an Egyptian-colloquial lexicon based on 380 sentiment words seeds, it was then expanded taking into consideration the patterns in which these seeds are existed.

[16] developed a gold-labelled dataset for Saudi dialect, the dataset is called "Saudi Twitter Corpus for Sentiment Analysis" which includes 4700 tweets., they have followed the footsteps of [15], and they've utilized patterns to expand their lexicon. Other terms from a [17] lexicon were added to the Saudi lexicon after normalization processes. Their lexicon consists of 14 k terms.

A recent study by [18] proposed an Arabic dataset that includes Tweets and Emojis. The authors produced this dataset using automated lexicon-based approach to annotate the dataset. This dataset initially includes six million tweets. In this study, the author addressed how to deal with multiple Arabic dialects. Moreover, various Arabic dialects were translated into Arabic MSA expect for Iraqi dialect.

There are some limitations in the traditional sentiment analysis which caused researchers to look after aspect-based SA while dealing with reviews that are related to various product aspect. [19] Claim that they have been the first to deal with aspect-based SA in Arabic, as they developed an annotated human Arabic Dataset (HAAD), consisting of the 1513 LABR Arabic book reviews. In their research they carried out four tasks: T1 (Aspect Term Extraction), T2 (Aspect Term Polarity), T3 (Aspect Category Identification), and T4 (Aspect Category Polarity). Seven groups of 3 students each made the annotation using a web-based annotation tool (BRAT). In the same year, [20] showed the effect of Arabic news on the readers, they collected data from social media and a well-recognized network of Arabic media like Al Arabiya and Al Jazeera throughout the war that occurred between Gaza and Israel in 2014. Over 10 k postings were gathered, 2265 of which were annotated with a supervised learning approach. During this experiment, two classification techniques were evaluated: CRF and J48. The results showed that in aspect-terms extraction J48 was better than CRF, while CRF outperformed J48 in aspect-term polarity identification. However, POS tags and NER types were an important part of assessing the news effect.

[21] Presented a baseline approach to Support Vector Machine (SVM) in Arabic hotel reviews. Out of 15,562 comments from popular websites of hotel reservations like (booking.com, tripadvisor.com) collected in several Arab cities, 2291 reviews were selected. Both, text level and sentence level annotations were applied, sentence was annotated based on SemEval-ABSA15 principles, and text was annotated based on SemEval-ABSA16 task 5.

Two supervised approaches to machine learning were suggested in line with task 5 of the SemEval-ABSA16 by [22], namely deep recurring neural network (RNN) and support vector machine (SVM). The authors investigated three different tasks as per SemEval: aspect category identification, aspect opinion target expression (OTE), and aspect sentiment polarity identification. In order to assess the overall effectiveness of the suggested approaches, SemEval-2016 review data set from Arabic Hotels was used. Compared to baseline research, the findings showed that SVM outperformed the deep RNN in all examined tasks. Though, while comparing the time consumed during training and testing, the deep RNN was discovered to be better and faster. Additionally, [23] suggested an improved approach of supervised machine learning which could extract aspects and then classify sentiments in Arabic hotel reviews in SemEval-2016 dataset. This approach covered three tasks: identify aspect categories, extract opinion targets, and identify the sentiment polarity. The evaluation results showed that their approach exceeded the benchmark criteria using the same dataset.

3 Dataset Construction

This section provides illustration of the approach followed to construct the domain specific dataset and related dictionaries. At our utmost knowledge, our dataset is the first resource which targets the Arabic government mobile app reviews. A manual annotation approach was followed due to the unstructured nature of the Arabic reviews which has both MSA and multiple dialects. According to [24], manual approaches are more efficient and produces more accurate datasets compared to other approaches. However, the only drawback for this approach that it consumes a massive amount of time and resources. Fig. 1 illustrates our approach in constructing the dataset.

3.1 Dataset Analysis

In this research, we adopted the dataset produced by [9]. The data set comprises of 12 k reviews. The source of these reviews is Apple Store and Google Play which are related to 60 different mobile apps for governments in United Arab Emirates (UAE). 2 k reviews from the 12 k were in Arabic Language. The Arabic part contains a mixture of MSA and multiple dialects, this could be due to the variety of Arab nationalities living in UAE. These reviews were posted between the years 2013 and 2016 when HH Sheikh Mohammed bin Rashid Al Maktoum announced the initiative

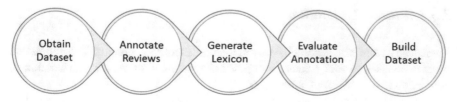

Fig. 1 Dataset construction approach

Table 1 Aspects examples

Aspect	Example 1	Example 2	Example 3
User Interface	تصميم رائع	البرنامج جميل	يحتاج ترتيب من ناحية الاقسام
User Experience	أفضل برنامج	خدمة مميزة ومريحة	برنامج فاشل
Functionality & Performance	برنامج سريع	يوفر جميع الخدمات	لا يدعم اللغة العربية
Security	طلبتم إذن الدخول الى الكاميرا وهذ يسمى اختراق خصوصية	برنامج يتجسس على كل شيء	احاول ادخل الحساب و مب راضي
Support & Update	رائع بعد التحديث	مش راضي يثبت، مع انه كان شغال قبل كدا	أرجو العمل على حل المشاكل التقنية

of "Smart Government". The data includes: the name of the app, name of the store, ID of the application, ID of the review, language, date, star rating, subject, body, and author. Our focus in this research was on the 2071 Arabic reviews which constitute 17% of the 11,912 reviews. Most of the reviews were comparatively rated positively. For instance, 1421 reviews were rated with five stars, 123 were rated as four stars, while 99 reviews were rated as three stars, 76 have got two stars, whereas 352 as one star. There was 60 dummy and spam reviews (2.9%) which have been eliminated. It has been noticed that reviews contain a lot of slangs, therefore many spelling mistakes found which required additional effort to fix them.

3.2 Aspects-Based Approach

It is essential to define the aspects or features that will be used to categorize the reviews [24]. For instance, the memory and screen resolution are different aspects of smart phones. As mentioned previously, this research is in continuation of previous work made by [9] which adopted aspect-based sentiment analysis. [9] Defined a set of aspects based on various standards such as Smart Website Excellence Model (SWEM), Apple iOS principles and guidelines, along with Google Android quality guidelines. These aspects are: *User Interface, User Experience, Functionality and Performance, Security, Support and Updates.* Table 1 illustrates theses five aspect terms with some examples on each aspect that have been extracted from the Arabic reviews:

3.3 Manual Annotation

There was a team who accomplished the manual annotation process. The annotation process consists of four primary sub processes: annotation, review, verification of annotation and approval of annotation.

A specially designed computer application named "GARSA" which was developed by [9] have been used in this study to complete the annotation process. The application enables the annotators to retrieve the reviews sequentially from a database that stores all reviews, then each review will be read, analyzed and annotated by marking the opinion word with the corresponding term which will be assigned to an appropriate aspect. After that, the annotators will double-check whether if the given star rating is matching the overall sentiment of the review based on their judgement after understanding the reviewer point of view. Once the annotators complete the annotation, another member will verify the annotation, and send it for final approval by the domain expert.

3.4 Generate Lexicon

A sentiment lexicon is an essential resource for trying to identify the polarity of mobile user reviews [25]. In order to generate our sentiment lexicon, after the final approval of domain expert on the annotation process, some additional tasks were applied on the data. It was necessary to clean out all diacritics and punctuation marks from the reviews, as shown in Table 2.

Then, by using a Python regular expression, the normalization process was executed in order to remove (Hamza - ء) and (Madda - ~) from letters like (Alif – ئ إ أ ا), in addition to converting (Taa Marbotta – ة) to become (ه) as shown in Table 3.

After that, the stemming task is executed on the resulted data produced from the previous task, to exclude suffixes such as (كم، ين، وا)and prefixes such as (بال هال، فال، ال)and all feminine pronouns such as (تها، ها)as shown in Table 4.

Finally, duplicated letters are removed, as shown in Table 5.

All opinion words with aspect terms will be stored in a special table inside the database in order to execute a function that will result in specifying the polarity score for each opinion word based on the final star rating by applying a formula that will assign a 5-stars rating review a numeric value of (1.0), 4-stars, 3-stars, 2-strars, and

Table 2 Diacritics and punctuation examples

Task	Before	After
Punctuation Removal	جميل.	جميل
	مناسب،	مناسب
	فاشل!	فاشل
Diacritics Removal	مُتقدّم	متقدم
	شكراً	شكرا
	مُمتاز	ممتاز

Table 3 Normalization process examples

Task	Before	After
Normalization Process	أجمل	اجمل
	إبداع	ابداع
	حلوة	حلوه

Table 4 Stemming examples

Task	Before	After
Stemming	بالابداع	ابداع
	مشكورين	مشكور
	انصحكم	انصح
	يحملها	يحمل

Table 5 Stemming examples

Task	Before	After
Duplicated Letters Removal	ر اااائع	رائع
	عجيبببب	عجيب
	كفوووو	كفو

Table 6 Sentiment word polarity examples

Sentiment Word	Polarity
ابتكار	1.0
يحتاج	-0.1875
ابداع	0.7857
يعلق	-0.75
جميل	0.9032
فاشل	-0.875

1-star will be assigned the numeric values (0.5, 0, −0.5, −1.0) respectively. Then it will calculate the number of iterations of each opinion word under each aspect, and multiply it with each star rating value. Finally, it will get the average by dividing the result on the total number of iterations as shown below:

$$PS = \frac{(1SRV \times ITE1) + (2SRV \times ITE2) \dots (5SRV \times ITE5)}{(ITE1 + ITE2 + \dots + ITE5)}$$

where:

- *PS is polarity score,*
- *SRV is star rate value,*
- *ITE is iteration.*

Based on the above equation, the polarity score will range from −1.0 (negative words) to 1.0 (positive words). The produced lexicon consists of 218 Arabic words, some of the Sentiment Words are shown in Table 6 with the calculated polarity:

3.5 Building the Dataset

As mentioned previously, GARSA was utilized to annotate 2071 reviews manually by identifying opinion words, with the corresponding term, then classify it with the

```
"4": {
    "review": "جهاز تحديد موقع العميل: تصميم موقع العميل: تصميم رائع
    تطبيق سهل ومفيد",
    "rating": "5",
    "aspects_ratings": "user interface:5 |
                        user experience:5 |
                        performance:0 |
                        security:0 |
                        support & update:0 |
                        Overall:5",
    "sentiment_word": [
        "رائع",
        "مفيد",
        "سهل"
    ],
    "term_name": [
        "تصميم",
        "تطبيق",
        "تطبيق"
    ],
    "polarity": [
        0.96078431372549,
        "0.870967741935483",
        1.0
    ],
    "aspect_name": [
        "User Interface",
        "User Experience",
        "User Experience"
    ]
```

Fig. 2 JSON format record

appropriate aspect class. Each review has an average of 7 words. The polarity score was calculated for each opinion word using a formula on the generated lexicon. The next step is to identify the overall rate for each aspect by calculating the average polarity score in each aspect class.

Finally, the dataset is built in JSON format containing a record for every review. The record information carries: review body and subject, rating, extracted opinions with the aspect class with calculated polarity as shown in Fig. 1 (Fig. 2).

3.6 Inter-Annotator Agreement Evaluation

Due to the fact that annotation process was conducted by a team. It was important to evaluate the precision of annotation between the team members. A random selection of 220 reviews was made to compare and observe the level of agreement between

Table 7 Aspect matrix of annotators agreement and disagreement

Aspects	Agreement and disagreement				K (%)
	TT	TF	FT	FF	
User Interface	101	2	2	11	82.67
User Experience	137	2	2	18	88.56
Functionality and Performance	113	3	5	25	82.79
Security	36	2	2	16	83.63
Support and Updates	63	4	1	19	84.58
Total	450	13	12	89	84.98

all team members' annotation. This evaluation will ensure that all members have a proper and common understanding of the guidelines and guarantees a high degree of annotation confidence among team members. The agreement metric gives an idea of areas of disagreement that may require further clarification for annotators. The statistics of Cohen's Kappa [26] are selected to measure the inter-annotator agreement. The following formula can define Kappa:

$$k \equiv \frac{Pa - Pc}{1 - Pc}$$

where:

- *Pa is the qualified agreement that has been spotted.*
- *Pc is the theoretical probability of a random agreement.*

Table 7 demonstrates the aspects matrix among annotator's agreement and disagreement. The team members agreed on the same aspect in 450 cases as shown in (TT), where they didn't agree in 89 cases (FF). On the other hand, some members agreed and some other disagreed in 13 and 12 cases as shown in (TF) and (FT) respectively. The results calculated for Cohen's Kappa (K) measures, are all greater than 81% [26]. Stated that if Kappa measures exceeds 81%, then the level of agreement is pretty perfect. This result confirms that all annotators could perceive a proper and common understanding of the guidelines and assures a high degree of annotation confidence among team members.

4 Aspects Extraction and Sentiment Classification

4.1 Task Definition

This section outlines the approach proposed for the development of an integrated lexicon and rule-based ABSA model. The rule-based ABSA model was chosen due to the unstructured nature of Arabic reviews which has both MSA and multiple

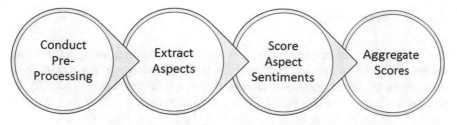

Fig. 3 Aspects extraction and sentiment classification approach

dialects. [27] Pointed out the importance of rule-based approach compared with other approaches since the rules are created manually which ease the integration between domain knowledge and NLP tasks. It extracts aspects and classifies the corresponding sentiments from government apps reviews in Arabic. In the next section, the approach of extracting explicit and implicit aspects is illustrated, besides the sentiment classification (Fig. 3).

4.2 Dataset

The model employed a lexicon that was manually generated in this study with hybrid rules, in particular to deal with part of key challenges in ABSA and SA as a whole. Table 8 summarizes content statistics of the dataset.

Table 8 Content statistics of the dataset

# of reviews	2071
# of characters (No Spaces)	64,869
# of characters (With Spaces)	77,980
# of Words	14,059
# of Explicit Aspects Terms	158
# of Implicit Opinion-Aspects Terms	193
# of Opinion Words	240
Average Words per Review	6.79
# of Very Positive Reviews	1422
# of Positive Reviews	124
# of Neutral Reviews	100
# of Negative Reviews	77
# of Very Negative Reviews	353

4.3   Pre-processing

It is crucial that text preprocessing tasks are performed when using a rule-based approach with supportive lexicons. Initially, the algorithm we propose is to divide each review based on the punctuation marks so it becomes easier to identify the phrases starting and ending points. Such punctuation marks: full-stop (.), question mark (?) and exclamation mark (!). This will significantly affect the link between the polarity score and the correct aspect term without intervening with the irrelevant phrases.

On the other hand, the subject of the review will be the first phrase in the review. Then, tokenization will be performed on the phrase so that each token that contains a punctuation mark will be excluded and letters will be transformed to lowercase. At this stage, the normalization task such as the removal of duplicated characters, will not be performed, because the scoring phase of aspect sentiment will be able to treat such cases and consider it intensification that impacts the scoring polarity. For instance, the word *"رائع/////"* which means *"wonderfulllll"* will be considered as *"extremely wonderful"*. Next, all stop words such as *(هذا) ,(من)* will be excluded based on a customized list of stop words which have been taken into consideration while reviewing the dataset initially.

4.4   Implicit and Explicit Aspect Extraction

As a matter of fact, to complete the aspect extraction task in SA, it is necessary to identify aspects categories [9]. Have specified the aspect categories based on the standards of Apple, Android, and Smart Dubai Government (SDG). The identified aspects categories are: *User Experience, User Interface, Security, Functionality and Performance, and Support and Updates*. In this study, the same aspects categories will be used.

There are many challenges in the aspect extraction task, but the key challenge is that aspects are not always mentioned explicitly in the reviews. In other words, an explicit phrase will contain a term which points directly to the aspect category. However, the case will be different while dealing with implicit phrases where no term can be found that indicates the aspect category, which makes the aspect extraction more complicated while dealing with implicit phrases.

For example, *"تصميم رائع، تطبيق سهل"* which means *"wonderful design, easy and useful app"*, in this case, the first part of the phrase has explicitly mentioned the term "design" with an opinion word *"wonderful"*. This results into a positive sentiment toward *"User Interface"* aspect category. On the other hand, there is no such an explicit term that points directly to the aspect category while dealing with the second part of the phrase *(Something like: "تطبيق سهل الاستخدام")*. However, the opinion word *"سهل"* implicitly points to *"User Experience"* aspect category.

[7] Stated that one of the ways that is vastly used in order to identify an aspect, is to deem opinion words as a high potential nominee for the extraction of the implicit aspect. Therefore, the algorithm will initially look for the opinion words that points to the aspect based on the lexicon. If it couldn't specify the aspect category, then it will look for the nearest aspect term in the same phrase, taking into consideration a maximum window size of two with higher priority to the left side given that in Arabic language, adjective follows the term. Finally, the identified words for opinion and aspect are examined by the lexicon to specify the category of the aspect.

4.5 Aspect Sentiment Scoring

The approach followed uses the established lexicon which was produced during this study. Practically, the algorithm will travel through the phrases and when an opinion word is recognized, it will retrieve the polarity score from the lexicon and associate it with the extracted aspect. Multiple settings have been applied during the experiment of the algorithm for the identification of opinion words within a phrase, apart from using the lexicons. For example, many rules for intensification, negation, downtoning, duplicated characters and the special cases of opinion negation rules have been adopted to be handled.

The basic baseline approach followed is by using a lexicon to directly lookup the polarity score of a sentiment word. But due to the fact that polarity accuracy can be affected with many factors, such as intensification, negation, and downtoning [28], this score has not considered to be final. A lexicon that indexes different terms implying negations has been created for each parameter of these modifiers. This means, the word "*not*" for example which implies negation, when it falls beside opinion word, the polarity score of the opinion word will be multiplied by (-1). This is called "Switch Negation" since it reverses the polarity in the opinion word. However, the algorithm can treat other cases of negation where the opinion word has a high positive strength, such as "تطبيق غير ابتكاري"which will not invert the polarity score, but it will reduce it by performing polarity shifting through multiplying the score with some certain threshold (e.g. 0.5) as an alternative of reversal. While shifting the negation, highly positive opinion words that have a score which is higher than (0.75), and highly negative opinion words that have a score less than (-0.75) can be identified through the proposed approach. Table 9 demonstrates some examples of the negation words from the negation's lexicon.

Intensifiers and downtoners are treated through the proposed algorithm as it can affect positively or negatively on the polarity score of an opinion word whenever they appear in the phrase [29]. Consequently, the proposed algorithm will detect any intensifier or downtoners based on the lexicon. Usually, when dealing with Modern Standard Arabic (MSA) intensifier or downtoners falls after the opinion word such as "رائع جدا". However, this may change while dealing with Dialect Arabic (DA), as intensifier or downtoners may fall before the opinion word such as "جدا جميل". Moreover, the impact of the intensifier or downtoners differs from one case to another.

Table 9 Negation Words Sample

Negation Term	Example
لا	البرنامج لا يعمل
ما	بصراحة ما يستاهل التثبيت
بدون	كل ميزات البرنامج تشتغل بدون مشاكل
لم	لم أتمكن من تنزيله بعد عدة محاولات
عدم	عدم استجابة التطبيق لأي من الأوامر
مش	شكل البرنامج مش حلو
مب	شو فايدة البرنامج يوم انه مب شغال عدل
غير	بعض المعلومات غير صحيحة

Therefore, a multiplication factor has been created for each case as shown in Table 10 which illustrates some intensifiers and downtoners. These intensifiers and downtoners were created during the annotation process through the experts' judgement, where the experts are native Arabic language specialists. During the annotation process, the experts categorized the intensifiers into very high (2.0), high (1.75), medium (1.5) and low (0.25). Likewise, downtoners were categorized into low (0.75), medium (0.5) and high (0.25). On the other hand, modifications may appear in various forms and in few cases, it can be incorporated into the opinion word itself. Some of these cases were treated through the lexicon such as "رائع" and "روع" where the second opinion has a higher polarity of the first one. Also, people may express their feelings by using repeated letters such as "تطبيق ممتاااااز" which means "*excellent application*". This can be treated as another form of intensifier. Similarly, [30] suggested to consider duplicated characters as an intensification and proposed a solution to improve the word sentiment marking both for positive and negative orientations whenever two or more consequent characters are found.

Table 10 Some Intensifiers (I) & Downtoners (D) Examples

Term	Type	Factor	Example
جدا	I	1.75	برنامج جميل جدا
بجد	I	1.50	روعة بجد
جد	I	1.50	عن جد واو
قمة	I	2.0	قمة الابداع
والله	I	1.25	سيء والله باقي البرامج أحسن
كتير	I	1.75	بصراحة كتير مفيد
بشدة	I	2.0	أنصح به بشدة
كل	I	1.25	كل الشكر على هذا الانجاز
وايد	I	1.75	وايد مفيد
تماما	I	1.50	لا يعمل معلق تماما
اكثر	I	1.75	مميزات التطبيق اكثر من رائعة
للغاية	I	2.0	سيء للغاية
بعض	D	0.25	يحتاج بعض التحسين
شوي	D	0.50	كثرة التنبيهات مزعجة شوي
نوعا ما	D	0.75	التحديث الاخير نوعا ما احسن
بعض الشيء	D	0.75	أنه بطيء بالاستجابة بعض الشيء
قليلة	D	0.25	في مشاكل قليلة مثل عدم إمكانية تعديل الاسم
صراحة	I	1.50	روعة صراحة
فوق	I	1.50	فوق الممتاز
احيانا	I	1.25	يحتاج تحديث احيانا حتى يشتغل

4.6 Aspect Sentiment Aggregation

The algorithm aims to specify the star rating for various aspects derived from the review. The star rating is determined during the experiment where: 1-star considered as very negative, 2-star as negative, 3-star as neutral, 4-star as positive, and 5-stars as very positive sentiment. This will give a detailed vision of the user's feedback as it can open the app owners' eyes on cons and pros of each aspect instead of having a general positive or negative feedback. As a result of this, opinion words will be extracted with the final polarity score that associated to the relevant aspect. Now, the calculation of the average polarity rating for opinion words which is categorized under each aspect would be a straight-forward procedure. The following illustrative example shown in Table 11 provides a scenario of a user review with the calculation of the final star rating for extracted aspects:

"شكل البرنامج جميل جداً وقمة بالترتيب والأناقة. بس لاحظت انه ما يشتغل بدون انترنت وفيو شوية مشاكل في تحديد الموقع"

The aggregation task classifies and calculates the average polarity by aspect category. Below is the explanation of the given examples in Table 10:

1. **User interface** $= \frac{1.68+1.26+1.84}{3} = 1.59$
   (*This is considered as a 5-star rating*)
2. **Functionality and Performance** $= \frac{(-0.36)+(-0.33)}{2} = -0.35$
   (*This is considered as a 2-star rating*)

The last step of ABSA is to summarize the whole series of reviews with a graphical representation using a data visualization that shows an equivalent star rating for each aspect on a five-star scale.

4.7 Experimental Evaluation

There were several attempts to improve the performance of the algorithm for aspects extraction and opinion classification. The evaluation of aspects extraction and sentiments orientation were performed based on various settings and parameters. To measure performance, standard confusion matrix is used. This can definitely support in calculating the advanced evaluation measures, namely Accuracy, Precision, Recall, and F-measure. The tables (Tables 12 and 13) illustrate all elements of the

Table 11 Calculated Average Sample

Opinion	Aspect	Score	Modified By	Final Polarity	Star Rating
جميل		0.9	1.75	1.68	
ترتيب	User Interface	0.63	2	1.26	5-Stars
أناقة		0.92	2	1.84	
يشتغل	Functionality &	0.36	-1	-0.36	
مشاكل	Performance	-0.65	0.5	-0.33	2-Stars

Table 12 Confusion Matrix (Aspect Extraction)

		Predicted (Proposed Aspect Extraction Approach)	
		Retrieved	Not Retrieved
Actual (Annotated Dataset)	Relevant	TP (true positive)	FP (false positive)
		Number of aspects that are correctly extracted	Number of aspects that are annotated, but the not extracted by the algorithm
	Irrelevant	FN (false negative)	TN (true negative)
		Number of aspects that are not annotated, but extracted by the algorithm	Number of aspects that are not annotated and not extracted by the algorithm

Table 13 Confusion Matrix (Sentiment Classification)

		Predicted (Proposed Sentiment Classification Approach)	
		Retrieved	Not Retrieved
Actual (Annotated Dataset)	Relevant	TP (true positive)	FP (false positive)
		Number of sentiment polarity scores that are calculated correctly by the algorithm	Number of sentiment polarity scores that are incorrectly calculated by the algorithm
	Irrelevant	FN (false negative)	TN (true negative)
		Number of aspects that does not have sentiment assigned, but calculated by the algorithm	Number of aspects that does not have sentiment assigned and not calculated by the algorithm

confusion matrix for aspect extraction and for sentiment classification:

The evaluation measures of the performance are calculated based on the following standard formulas:

1. $Accuracy = \frac{TP+TN}{TP+TN+FP+FN}$
2. $Precision = \frac{TP}{TP+FP}$
3. $Recall = \frac{TP}{TP+FN}$
4. $F-measure = \frac{2*Precision*Recall}{Precision+Recall}$

5 Results Discussion

The results for aspect extraction versus sentiment classification are shown in Table 14. It represents a comparison between the experiments that have been carried out during this study, keeping in mind that aspect extraction has taken into consideration both implicit and explicit aspects.

Meanwhile, some misleading aspect term showed up beside the recognized opinion word, where in this scenario, sentiments have been allocated false aspect and correct aspect could not be extracted. This will increase the number of FP and FN and result in negatively impact the Precision and Recall measures. Table 15 shows some examples:

As mentioned in Sect. 4.5, multiple settings of parameter have been used to measure the improvement progress of ABSA. Initially, the basic approach considers lexicon (LEX) as a baseline without taking into consideration any other rules. Then, additional rules settings have been added to measure the performance improvement. The purpose of adding these rules settings and adopt them in the algorithm is to handle some of the challenges in SA such as: dealing with negation (NEG), handling intensifiers (INT), downtoners (DWT), duplicated characters (CHR) and special cases of negation-opinion rules (SCR). The improvement in the progress is presented in (Table 16), where best performance in term of aspect extraction and opinion classification could be achieved when implementing all rules settings i.e. (LEX + NEG + INT + DWT + CHR + SCR).

Table 14 Performance Evaluation Results (Aspect Extraction vs. Sentiment Classification)

	Performance evaluation			
	Precision (%)	Recall (%)	F-Measure (%)	Accuracy (%)
Aspect Extraction	90.54	94.55	92.50	96.57
Sentiment Classification	90.22	91.15	90.68	95.81

Table 15 Misclassified Reviews Sample

Review	Aspect Term	Opinion Word	Aspect Category	
			Extracted Aspect	**Correct Aspect**
بصراحة تميز في استخدام مجموعة متنوعة ومتناسقة من التصاميم في البرنامج	التصاميم	تميز	User Experience	User Interface
تعجبني الأشكال المتاحة والآمنة لعملية الدفع	الدفع	تعجبني	User Interface	Security & Update
روعة في التصميم وراحة أثناء القيام بمختلف المميزات المتاحة	المميزات	راحة	User Interface	Functionality & Performance

Table 16 Performance Improvement Results

		Performance evaluation			
		Precision (%)	Recall (%)	F-Measure (%)	Accuracy (%)
Parameters	LEX (Baseline)	68.24	80.00	73.65	89.47
	LEX + NEG	75.46	82.55	78.85	91.62
	LEX + NEG + INT	79.03	85.23	82.01	92.74
	LEX + NEG + INT + DWT	81.49	87.11	84.20	93.63
	LEX + NEG + INT + DWT + CHR	82.99	87.56	85.22	93.98
	LEX + NEG + INT + DWT + CHR + SCR	**90.22**	**91.15**	**90.68**	**95.81**

6 Theoretical and Practical Implications

As a result of this experiment, many theoretical and practical implications could be identified. Theoretical implications such as addressing various SA challenges in Arabic language which can improve performance including the extraction of implicit aspects, handling all types of sentiment modifiers, such as negations, intensifiers, and downtoners. With an enhanced list of negations, intensifiers and downtoners, the F-message have a significant improvement by 11% and the accuracy by 4% in comparison with the lexical baseline, respectively. On the other hand, [6] stated that the aspects are not clearly defined in terms of implicit aspects and that the sentence semantics indicates implicitly the aspect. In the Arabic reviews of government's mobile apps domain, considering implicit aspects can be essential as it can create a clear understanding of the review and how users can express it, which is commonly used in many domains.

Because of the rapid development in technology, companies need to listen effectively to the public's voice to improve their products or services which results in attracting customers for using those products or services. Global strategies direct businesses to focus on customers and design their services according to their customers' needs. The proposed model can play a critical role in understanding the view point of Arabic citizens toward the services provided by government. As it can help in identifying the weaknesses and strength points of the provided services, which will result in enhancing services that don't meet the customers' expectations in order to offer better services that will retain customers and keep them satisfied.

7 Conclusion

This research targeted a new untouched domain in ABSA. We proposed a combined approach to extract aspects and classify sentiments from Arabic government apps Arabic reviews by adopting lexicon-based and rule-based approaches. The proposed approach used a manually annotated dataset in this study for government mobile apps Arabic reviews. In this research, various challenges were tackled to improve the performance of aspect extraction and sentiment classification. For instance, these challenges include extracting explicit and implicit aspects, dealing with negations, handling intensifiers and downtoners, duplicated characters among others. The performance evaluation showed significant results in extracting aspects and classifying sentiments. For instance, combining both explicit and implicit aspects extraction approach have achieved significant performance in terms of Accuracy and F-Measure by 96.57% and 92.50% respectively. In terms of sentiment classification, the results reported in the study have shown an increase in the accuracy and f-measure by 6%, and 17% respectively when compared with the baseline. This model can help government organizations in identifying their clients' needs and expectations through scientific approach by studying their feedback and reviews on their government applications and services. Moreover, this approach can provide an important input to identify potential improvement areas. This increases the client's satisfaction and happiness and aligns with Dubai Governments objective in making Dubai the happiest place on earth.

References

1. M.Z. Asghar, A. Khan, S. Ahmad, F.M. Kundi, A review of feature extraction in sentiment analysis **4**(3), 181–186 (2014)
2. S.Y. Ganeshbhai, *Feature Based Opinion Mining : A Survey* (2015), pp. 919–923
3. B. Liu, L. Zhang, *A Survey of Opinion Mining and Sentiment Analysis* (2012)
4. M. Hu, B. Liu, *Mining and Summarizing Customer Reviews* (2004)
5. G. Carenini, R. Ng, A. Pauls, Multi-document summarization of evaluative text. Comput. Intell., 305–312 (2013)
6. S. Poria, E. Cambria, A. Gelbukh, Aspect extraction for opinion mining with a deep convolutional neural network. Knowl.-Based Syst. **108**, 42–49 (2016)
7. S. Poria, E. Cambria, L.-W. Ku, C. Gui, A. Gelbukh, A rule-based approach to aspect extraction from product reviews, in *Proceedings of the Second Workshop on Natural Language Processing for Social Media (SocialNLP)* (2014), pp. 28–37
8. M. Tubishat, N. Idris, M.A.M. Abushariah, Implicit aspect extraction in sentiment analysis: review, taxonomy, oppportunities, and open challenges. Inf. Process. Manage. **54**(4), 545–563 (2018)
9. O. Alqaryouti, N. Siyam, K. Shaalan, *A Sentiment Analysis Lexical Resource and Dataset for Government Smart Apps Domain*, vol 639 (Springer International Publishing, 2018)
10. J. Moreno-Garcia, J. Rosado, Using syntactic analysis to enhance aspect based sentiment analysis, in *International Conference on Information Processing and Management of Uncertainty in Knowledge-Based Systems* (2018), pp. 671–682
11. C.C. Aggarwal, C. Zhai, *Mining Text Data*, vol 101, no 23 (2012)

12. M. Rushdi-Saleh, M.T. Martín-Valdivia, L.A. Ureña-López, J.M. Perea-Ortega, OCA: Opinion Corpus for Arabic, *In vivo (Athens, Greece)*, vol 30, no 2 (2011), pp. 155–157

13. M. Abdul-Mageed, M.T. Diab, M. Korayem, Subjectivity and sentiment analysis of modern standard Arabic. Assoc. Comput. Linguist. **29**(3), 587–591 (2011)

14. M. Abdul-Mageed, M.M. Diab, AWATIF: a multi-genre corpus for modern standard arabic subjectivity and sentiment analysis, in *Language Resources and Evaluation Conference (LREC'12), Istanbul* (2012), pp. 3907–3914

15. S.R. El-beltagy, A. Ali, *Open Issues in the Sentiment Analysis of Arabic Social Media : A Case Study*, no. June, 2013

16. A. Assiri, A. Emam, H. Al-Dossari, Towards enhancement of a lexicon-based approach for Saudi dialect sentiment analysis. J. Inf. Sci. **44**(2), 184–202 (2018)

17. G. Badaro, R. Baly, H. Hajj, A large scale Arabic sentiment lexicon for Arabic opinion mining, in *Arabic Natural Language Processing Workshop Co-located with EMNLP 2014* (2014), pp. 165–173

18. H. Abdellaoui, M. Zrigui, *Using Tweets and Emojis to Build TEAD : an Arabic Dataset for Sentiment Analysis*, vol 22, no 3 (2018), pp. 777–786

19. M. AL-Smadi, O. Qawasmeh, B. Talafha, M. Quwaider, *Human Annotated Arabic Dataset of Book Reviews for Aspect Based Sentiment Analysis* (2015), pp. 726–730

20. M. AL-Smadi, M. Al-Ayyoub, H. Al-Sarhan, Y. Jararweh, *Using Aspect-Based Sentiment Analysis to Evaluate Arabic News Affect on Readers* (2015)

21. M. Al-smadi, O. Qwasmeh, B. Talafha, M. Al-ayyoub, Y. Jararweh, E. Benkhelifa, *An Enhanced Framework for Aspect-Based Sentiment Analysis of Hotels' Reviews : Arabic Reviews Case Study* (2016), pp. 98–103

22. M. Al-smadi, O. Qawasmeh, M. Al-ayyoub, Y. Jararweh, B. Gupta, Deep recurrent neural network vs. support vector machine for aspect-based sentiment analysis of Arabic Hotels' reviews. J. Comput. Sci. (2017)

23. M. Al-smadi, M. Al-ayyoub, Y. Jararweh, O. Qawasmeh, Enhancing aspect-based sentiment analysis of Arabic Hotels' reviews using morphological, syntactic and semantic features, in *Information Processing and Management*, no. October 2016 (2018), pp. 0–1

24. B. Liu, Sentiment Analysis and Opinion Mining Morgan & Claypool Publishers, in *Language Arts & Disciplines*, no. May (2012), p. 167

25. G. Badaro, R. Baly, R. Akel, L. Fayad, J. Khairallah, *A Light Lexicon-based Mobile Application for Sentiment Mining of Arabic Tweets* (2015), pp. 18–25

26. P. Takala, P. Malo, A. Sinha, O. Ahlgren, Gold-standard for topic-specific sentiment analysis of economic texts, in *Proceedings of the Language Resources and Evaluation Conference* (2010), pp. 2152–2157

27. K. Shaalan, *Rule-based Approach in Arabic Natural Language Processing*, no. May, 2010

28. D. Vilares, C. Gómez-Rodríguez, M.A. Alonso, Universal, Unsupervised (Rule-Based), Uncovered Sentiment Analysis ∗. Knowl.-Based Syst. **118**, 45–55 (2017)

29. M. Taboada, J. Brooke, M. Tofiloski, K. Voll, M. Stede, Lexicon-based methods for sentiment analysis. Assoc. Comput. Linguist. (2011)

30. F.M. Kundi, A. Khan, S. Ahmad, M.Z. Asghar, Lexicon-based sentiment analysis in the social web. J. Basic Appl. Sci. Res., 238–248 (2014)

Prediction of the Engagement Rate on Algerian Dialect Facebook Pages

Chayma Zatout, Ahmed Guessoum, Chemseddine Neche and Amina Daoud

Abstract For the purposes of online marketing, some social networks provide an advertising platform that allows the sponsoring of advertising content to reach target users. This content promotion is expensive in terms of the budget to be spent and this is why the content to be sponsored must be carefully selected. In other words, a company would ideally only sponsor content which is likely to perform well. The performance of an advertising content is usually measured by a metric called the Engagement Rate often used in the field of social media marketing to measure the extent to which the users will show "interest" for and interact with the advertised content. Thus, being able to predict the engagement rate of a publication is of utmost importance to Social Marketers. In this work, we propose a deep-learning-based system, to predict the performance of Facebook posts content in the Algerian Dialect, as measured by the users' engagement rate with respect to these publications. In order to predict the engagement rate, the system processes all the publication content: the text, the images, and videos if they exist. The images are preprocessed to extract their features and the Algerian Dialect textual content of the posts is analyzed despite its complexity which is due to multilingualism (use of Arabic, Algerian dialect, French and English). Two models of neural networks were proposed, one based on an MLP architecture and the other on a hybrid Convolutional-LSTM and MLP architecture. The results produced by these models on the prediction of the engagement rate are compared and discussed.

Keywords Social networks · Engagement rate · Arabic natural language processing · Algerian dialect · Image analysis · Neural networks · Deep learning · LSTM · Convolutional neural networks

C. Zatout · C. Neche
Department of Computer Science, USTHB, Algiers, Algeria

A. Guessoum (✉)
"NLP, Machine Learning, and Applications" Research Group, Laboratory for Research in Artificial Intelligence, USTHB, Algiers, Algeria
e-mail: aguessoum@usthb.dz

A. Daoud
SENSE Conseils, Hydra, Algeria

© Springer Nature Switzerland AG 2020
M. Abd Elaziz et al. (eds.), *Recent Advances in NLP: The Case of Arabic Language*, Studies in Computational Intelligence 874,
https://doi.org/10.1007/978-3-030-34614-0_9

1 Introduction

Social networks have undoubtedly become a primary destination for Internet users. The latter use social networks to achieve various goals, including getting up-to-date news of interest, keeping in touch with friends and acquaintances, and following their favourite brands and products. Social network sites are gradually becoming the most visited Internet sites. Indeed, in July 2018 the number of Internet users in the world reached 4.1 billion, of which 80.48% were active on social networks [1]. The most used social networks in the same month were Facebook with more than 2 billion users, followed by Youtube, WhatsApp, and Facebook Messenger with more than one billion users each [2]. The users' access to and interactions via social networks has generated a massive amount of data about their habits and preferences. This data is a gold mine that can be exploited for online marketing.

Thus, given this huge, world-wide popularity of social networks and the large amounts of data generated through the users' interactions on them, it has become a common practice for companies to adopt strategies for communicating with their audiences and for online marketing via social networks. Indeed, for the purposes of online marketing, some social networks provide an advertising platform that allows the sponsoring of advertising content to reach target users. This content promotion is expensive in terms of the budget to spend and this is why the content to be sponsored must be carefully selected. In other words, a company would ideally only sponsor content which is likely to perform well. Indeed, due to the significant costs of advertising on social networks, it has become essential for companies that decide to spend large amounts on advertising campaigns to publicize their brands and boost sales of their products, to monitor and measure the performance of these advertising campaigns. The performance of an advertising content is usually measured by a metric called the Engagement Rate often used in the field of social media marketing to measure the extent to which the users will show "interest" for and interact with the advertised content. Each platform offers its own formula for calculating the engagement rate. In general, the engagement rate represents the subscriber's engagement with regard to advertising content (such as posts or publications). The latter is characterized not only by its content (textual and/or visual characteristics) but also by other parameters such as the time of publication. These characteristics have a direct or indirect impact on the engagement rate.

The prediction of the engagement rate based on content characteristics can bring a real advantage to online marketers in that they can anticipate the engagement rate with respect to a post or publication and hence modify the content so as to raise the (predicted) engagement rate as much as possible. The idea is to use Machine

Learning techniques to process content characteristics and learn an engagement rate prediction model. Since these characteristics consist of images, text, and hyper-data, one machine learning approach that has proven extremely successful in visual object recognition, natural language processing, speech recognition, and many others, is Deep Learning.

Computer vision is one of the areas where deep learning has had most impact bringing noticeable advances in the produced results. Various architectures have been designed to meet the needs of computer vision problems. In addition, several challenges have been put in place each year to improve the best results obtained in the previous challenges and in various fields of applications. This is the case of ILSVRC, the ImageNet Large Scale Visual Recognition Challenge organized by ImageNet [3]. For the last few years, all the winners at this challenge used deep learning architectures: GoogleNet the winner at ILSVRC 2014, VGG-16 the second winning network in ILSVRC 2014; ResNet the winner at ILSVRC 2015; CUImage the winner at ILSVRC 2016 and BDAT the winner at ILSVRC 2017 [3].

We propose in this paper a deep learning system for the prediction of the engagement rate of Facebook publication content written in Algerian dialect. The challenge in our case is twofold: (1) we need to handle Algerian dialect publications and posts, a very complex task since the users tend to mix Modern Standard Arabic, French, English, and Amazigh words in their writings on social networks, in addition to the use of Arabic letters and/or Latin letters; and (2) develop a predictive model based on the text, (main) image and other content metadata.

The remainder of this paper is organized as follows. In Sect. 2, we discuss the difficulties encountered in processing the Algerian dialect. In Sect. 3, we present some related works. Section 4 presents the different steps followed in implementing our prediction system starting with data preparation and ending with the model design. In Sect. 5, we present and discuss the obtained results. A conclusion is given in Sect. 6 along with a presentation of some directions for future work.

2   The Algerian Dialect

This section is largely based on [4].

2.1   Generalities About the Algerian Dialect

In this paper, we use the expression "Algerian Dialect" to refer to the dialect spoken in Algeria. Its words are mostly derived from the Arabic language, though it has also borrowed many words from French, which are then conjugated as if they were Arabic words (verb derivations, plurals of nouns, etc.), and, to a lesser extent, from Berber,[1] Turkish, and rare words derived from Spanish [5]. This Algerian dialect, though fairly close to the Moroccan and Tunisian dialects, is quite easily distinguishable from each of them as far as its word morphology and some of the syntactic constructs; it is quite different from the Egyptian and other middle-eastern languages, though, by and large, any two Arabs, each one speaking his/her own dialect, would be able to understand each other, sometimes through some effort and provided each one speaks slowly enough.

To be more precise, there is not a unique Algerian dialect: various regions in Algeria speak slight variations of the same dialect with different accents on pronunciation. As such, the dialect from the eastern city of Annaba sounds a little different from that spoken in Algiers, the Capital which is located in the center north of the country, and both would sound different from the dialect spoken in the cities of Oran or Tlemcen in the Western region. Nevertheless, when used on social media, all these slight variations can broadly be treated as one, since all Algerians interact on social media using what is called "the Algerian dialect".

One finds in [6] a detailed comparison of the phonological, morphological, and orthographic differences between the Algerian dialect and the Egyptian and Tunisian dialects and MSA. [5] gives a brief presentation of the characteristics of the dialect spoken in Annaba (East of Algeria). We use them here and adjust them to the specificities of the "Algerian dialect". Here are some cases:

- Replacement of the diacritic kasra ("i" vowel/sound as in liberty) by the diacritic sukun (empty sound) at the beginning of a word (as in كْتَاب, "ktAb", instead of كِتَاب, "kitAb", in MSA);
- Bypassing or avoidance of the Hamza to be pronounced as a "ya" sound (e.g. "3A'ilah", family, is uttered "3Aylah");
- Some pronouns are slightly changed from their MSA form. E.g. نتوما("ntouma") for أنتم("antoum"), نْتِيَ(«ntiya») or نْتِ(«nti») instead of أنتِ(«anti»);
- For the possessive, it is common to add the word نتاعي(my) instead of just the MSA suffixed pronoun. For instance, one would equivalently say كْتَابي(«ktAby") or الكْتَاب نتاعي(al-ktAb ntA3y) to say «my book»;
- Some of the interrogative particles are slightly modified. E.g. وين("wyn" for "where"), شكون("shkoun" for "who"), etc.;
- The negation is introduced by means of the word ما(«mA») and the suffix شwith the diacritic sukun on it («sh») as in ماكليتْش(mA klItsh, for "I did not eat"). Negation also has some other expressions such as مارانيش(mA rAnysh, "I am not") or ماراناش(mA rAnAsh, "We are not"), etc.;
- As mentioned above, a number of words that are not of Arabic origin have found refuge in the Algerian dialect, such as طابلة("TAblah" for table, from French), كارْطابْلي("cArTably" for "my schoolbag", from French), etc.
- Most often, words of non-Arabic origin are conjugated using the rules applied to those of Arabic origin. For instance, طابلة("TAblah" for Table) gets a plural form as

---

[1]Berber is the language that was spoken in Algeria and other parts of North Africa before the Muslims from the Arabic Peninsula conquered North Africa.

طاءبلات("TAblAt" for tables) following the regular plural of Arabic feminine nouns (which is the form used for all nouns of foreign origin). Verbs of foreign origin are also conjugated as if they were of Arabic origin. For instance, مافريناتش("mA frynAtsh", "she did not pull the break").

The above, especially the part related to the use of words of non-Arabic origin, already highlights some of the challenges of processing the Algerian dialect.

2.2 The Algerian Dialect on Facebook

On the social media, Algerian users, hereafter Facebookers[2], use the Algerian dialect as well as French and English. This poses a major problem known as "Code Switching", that is to say using several languages and dialects within the same document, even sentence. We can thus find sentences like: "Bonjour, كيف حالك اليوم Brother?" (Good morning, how are you today brother?) where French, Arabic, and English are used in sequence in the same sentence. In addition, each region in Algeria has its own variations of the dialect, or even accent, which can affect written words. For example, the sentence قال لي("He told me") may be found written in different ways depending on the Algerian region "قالي" ("qAlly"), "الي" ("Ally"), "ڨالي" ("gAlly"), etc.

On the other hand, Algerian Facebookers may use the Latin alphabet to write Arabic words and phrases. For example, the sentence "كيف هي حالة الطقس"("how is the weather?") could be found as "kayf hiya HAlat elTaqs?" in Latin letters. One can also find the Arabic alphabet which gets used to write French words or phrases. For example, "Bonjour!" ("Good morning") may be found as "بونجور".

The absence of spelling rules has rendered the textual content of social networks characterized by intense orthographic heterogeneity: a word can be written in so many ways, based on the users' preferences or even habits. For instance, when Latin letters are used, 'mafihach' (meaning, "no it does not apply" or "it is not the case", depending on the context) can be written 'mafihech'. Moreover, a word written with Latin letters can contain numbers that represent some Arabic letters that do not exist as Latin letters; the word 'معلبليش'(I do not know), for example, can be written 'ma3labalich' since '3' is often used on social networks for the letter 'ع'. Even when Arabic letters are used, Algerian Facebookers tend to write words differently: 'معلبليش'(I do not know) can also be found as 'معلباليش','ما علاباليش', etc., let alone if Latin letters are used. Furthermore, some Latin transliterations of Algerian Dialect words have been found as also being English words. This is the case for 'وين'(where) transliterated as 'win' or as 'wine'. This is added complexity since the analyser of Algerian Dialect is supposed to recognize foreign words (French, English, mainly).

Another fairly common phenomenon on social networks is the repetition of letters in a given word. For example, the word "3laaaaaaaach" is equivalent to "3lach" ("why"). An inspection of this phenomenon [7] has suggested that often

---

letter repetitions are used to emulate spoken nonverbal cues and emphasise some utterance/statement.

As mentioned above, the fact that the Algerian dialect has its origins in formal Arabic with added vocabulary from Berber, Turkish, French, and Spanish makes it difficult enough to process. If, especially on social networks, we also consider the problem of code switching, with various possible spelling variations, writing errors and newly coined words, it goes without saying that this makes the Algerian dialect quite difficult to understand for the non-native user and very complex to automatically process.

3 State of the Art

With the growth of the number of Internet users, social networks have become a focus of large numbers of researchers, developers and even companies. Falcon.io [8] and Adobe Social [9], for example, are social media marketing platforms. They offer several marketing services for different social networks like Facebook, Twitter, Instagram, etc. They enable audience monitoring, sharing effective content, monitoring their performance and other activities.

One can find in the literature two types of content evaluation which can be based on content popularity or engagement. Although the two tasks are different, the techniques used in them are the same. For example, in [10] a video popularity is estimated based on clicks on it while engagement with respect to its content is related to the watching pattern. The latter evaluation can be based on users' profiles [11–13], content features [10, 14] or both [15]. It should be noted however that getting users' details and their centres of interests may be difficult or even impossible in some social networks where user privacy is maintained.

Chen et al. [11] have developed a model for predicting the engagement rate of a user according to the content of a post on Twitter. Indeed, it was to classify the impact of a post on Twitter in three categories: low impact, medium impact and high impact. To do this, they proposed three neural network techniques (a classic LSTM-RNN, a Bi-Directional Attention Model and a Convolutional 1d AlexNet). Chen et al. [15] have designed a model to predict the number of likes in Facebook content. For this, they combined two approaches: prediction based on the content of the publication and prediction based on the engagement of the user. The first approach consists in calculating the similarity between the content of the documents on the assumption that a user likes documents that are similar. The second approach calculates the similarity between users on the assumption that users with similar characteristics tend to like the same publications. Hu et al. [12] built a model based on the linear regression algorithm; it predicts the user engagement on Twitter with Real-World Events. In order to ensure this, they used tweet activities (such as total number of tweets, maximum tweets per hour, average tweets per hour and hashtag usage), tweet content (topical interests from tweet content and topical interests from the person's following list), geolocation (geographical proximity) and social network structure (such as number of followers); they mapped these into 17 numeric variables. Chamberlain et al. [13] designed a system that predicts the customer lifetime value in the network. Indeed, depending on the customers' purchases and the items these return, the authors' model predicts the loyalty (engagement) of a customer in a period of time. To do this, a customer is represented by a set of characteristics: number of purchases, number of sales, country, age, etc. Then machine learning techniques (Word Embeddings and Random Forest) were used to extract the relationships between the characteristics and to make the prediction. In [16], the authors adopted Latent Dirichlet Allocation [17], a probabilistic model to predict a user's engagement with online news. News text pieces were represented by a vector of words co-occurrences. The users' engagement was estimated taking into account the news piece textual content as well as the images and videos it contains.

As for image processing, [14] introduced a model based on an off-the-shelf Support Vector Machine for image popularity prediction. An image popularity is represented by its number of views on Flickr. For each image, a set of features is retrieved, namely: visual sentiment features, visual content and image context. The visual sentiment features are retrieved by finetuning DeepSentiBank [18], a deep convolutional network that classifies images into adjective-noun pairs. The visual features were extracted from another deep convolutional network [19]. As for an image context, the tags and image description provided with the image were used. Unfortunately, on some social networks like Facebook, tags and image descriptions are often not available. Moreover, when dealing with images used for advertising, as in our case, the image sentiment is always positive and thus retrieving it will be useless. However various researchers [20, 21] have confirmed that the colors used in an advertisement can have a positive effect on the targeted person. As a marketing tool, "color can be a sublimely persuasive force" [22]. This was our intuition when we decided to consider the colors used in a page as an additional feature that should affect a page popularity, and was indeed worth exploring.

Videos engagement is also an important aspect. In [10], the authors designed a linear regression model with L2-regularization to predict some engagement metrics such as average watch time of and relative engagement with YouTube online videos. In their case, videos were represented by a set of features like video duration, video category, language and channel activity level. However, the visual content of the videos was not explored even though it is an important feature.

In this paper, we propose a system for the prediction of social media content engagement. Since a post on social media can include text, images and videos which contribute to the quality of the post, hence the potential engagement of users with it, we explore in our work the use of text features, visual image features and visual video features to predict content engagement.

4   The Framework

Social networks generate a massive amount of data as a result of daily interactions between their users. The generated data are not homogeneous or fully structured. For example, a Facebook post consists of images and/or videos, a message, or both. The message itself is not structured; it may be written as a mixture of different languages and may contain emojis, links and hashtags. In addition, a Facebook post is characterized by other parameters such as the date of publication. This makes the exploitation of Facebook posts tricky.

We present in the sequel the design of a model for predicting users' engagement rate for Algerian brand posts. In order to build the model, we first collected the necessary data and then went to the various pre-processing steps such as cleaning up the data, transforming, reducing and discretizing it. Once all the pre-processing was applied, deep learning techniques were used and their performances compared so we could determine the most appropriate learning model.

Figure 1 illustrates the development and use of the system we propose for predict-

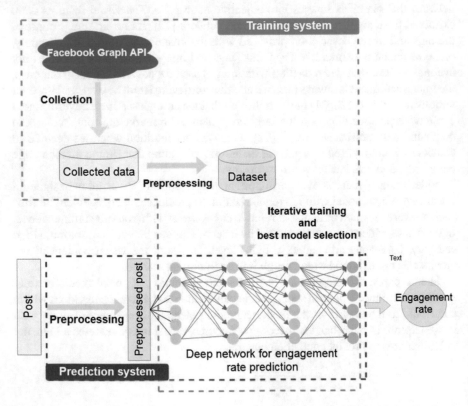

Fig. 1 System for predicting the users' engagement rate with respect to Algerian dialect Facebook posts

ing users' engagement rate with respect to Algerian dialect Facebook posts, using deep learning techniques.

The proposed system was developed in two phases: the training phase and the prediction phase. During the training phase, the focus was on building the prediction neural network model. This phase includes the different steps of data preparation and pre-processing, as well as training using different deep learning techniques. Once the model was built using the deep learning technique that yielded the best results, the prediction phase runs the developed model on new posts, after pre-processing them (following the same pre-processing steps as those of the training system), to predict their engagement rate. In this section, we detail each step we have followed to implement the engagement rate prediction system.

4.1 Data Preprocessing

4.1.1 Data Collection

The Facebook posts, 21,859 posts in total, were collected from 23 Algerian brand pages like 'Ooredoo Algeria', 'Nescafé', 'CevitalCulinaire' and others. We retrieved the page descriptions including the identifier, the number of subscribers and the list of all publications since the creation of each page. Moreover, we have downloaded for each post its components, such as the text, images and videos if they exist, as well as general characteristics like identifier, type, promotion status and creation date. We have also collected the parameters used to calculate the engagement rate. These are the number of reactions (including "Like", "Love", "Haha", "Wow", "Sad", "Angry"), the number of comments and the number of shares. We have used additional parameters that characterize an image, namely its identifier, type, location, width, length and, for the case of a video, its duration. We point out that for the videos we have only downloaded the image that comes first to perception.

4.1.2 Data Cleaning

We have removed publications which had a lifetime of less than 72 h since the engagement rate of these posts can change in an unpredictable way. Indeed, during the first 72 h of a publication lifetime, Facebook always sends them to the subscribers. We also removed the posts not affected by the engagement rate such as the polls. So we have removed from the dataset all posts having a type 'event', 'note' or 'offer'.

4.1.3 Attribute Extraction

Engagement rate: In the case of Facebook, the engagement rate of a publication is calculated in two ways depending on the available data and the purpose of the analysis: (1) for publication quality evaluation, and this requires access to the insights private page; or (2) for competitive analysis, that is to say to compare the performance between the publication of a brand and the publication of a competing brand. Since we did not have access to the pages private data, we used the following formula that uses public data only:

$$\text{Engagement rate} = \frac{\#Reactions + \#Comments + \#Shares}{\#Fans}$$

Note that the distribution of the engagement rate (Fig. 2) is not uniform over its domain. Indeed, its summary in five numbers reveals the following: the minimum value is 0; the maximum value is 2.31934; 25% of the publications have an engagement of less than 0.00006; 75% of publications have an engagement of less than 0.00086; and the average value of the engagements is 0.00019. In summary, engagement rate values have a mean of 0.00019 and a standard deviation of 0.02340.

General characteristics of a post: From the creation date we have extracted the month, the day, the day of the week, and the hour of publication.

Characteristics of the image of a post: From all the images contained in a post, we have decided to use only the main image. We have so decided because the main image usually has a greater resolution than the others and generally reflects the central theme of the page. As such we have analyzed that it attracts the subscribers' attention first. Nevertheless, to also take into account the fact that there may also be other images in the post, we have created the attribute "number of images". From the selected image, we have extracted its length, width, duration, and type. We have also built the following parameters: image shape (rectangle or square), the dominant colors and the vector of characteristics. We have decided that an image will be considered square if the difference between its length and its width does not exceed 10% of the length. The 10% threshold has been set empirically as the limit beyond which the human eye can perceive the difference between a square image and a rectangular one.

For the extraction of the dominant colors, we have tested and compared two different approaches. The first is to discretize the color attribute, that is to say to

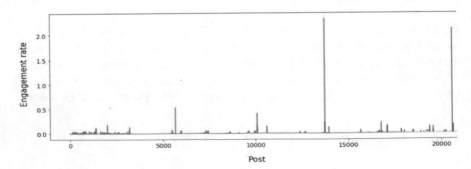

Fig. 2 Engagement rates of different posts

limit the number of bits representing the colors. For example, to reduce the number of possible colors to 8, the number of bits must be limited to three. The second approach uses the K-Means clustering algorithm [23] where k represents the number of dominant colors which has empirically been fixed to 16 colors. The first approach is faster in terms of execution time, but it is less accurate than the second. We opted for the second approach after comparing the distance between the dominant colors found and the image. In the following example (Fig. 3), the extraction of dominant colors using the K-means algorithm (Fig. 5) gives a better result compared to the quantization-based approach [24] (Fig. 4).

To represent the content of the image, we proceeded as done in [25]. We took the output of the last layer of VGG-16 and considered it as the set of visual features to use. The output of this layer is a vector of real size 1000.

Characteristics of the message of a post: Text pre-processing is a delicate phase. The pre-processing of a content of an Algerian publication is even more delicate. As mentioned in Sect. 2, messages written in Algerian dialect suffer from the problems of using different languages in the same message, composed with the use of Arabic and Latin characters.

Fig. 3 Sample image taken from the data set

Fig. 4 Dominant colors obtained using the quantification approach

Fig. 5 Dominant colors obtained using the K-Means algorithm

In order to represent the textual content of a post, in our case in a vector format, we have used three dictionaries: a dictionary for Arabic, one for French and one for English. The first hurdle indeed was to detect which language is used in each part of a post. The Arabic dictionary contains 9 million words. These words were automatically generated by the open-source AraComplex finite state transducer and validated using the 'Microsoft Word' spelling checker [26]. The French dictionary contains 208,000 words and the English dictionary contains 194,000 words. The last two dictionaries have been downloaded from the Gwicks site.[3]

Text normalization is an important operation that is performed to ensure that different word occurrences are used in a standard (canonical) form, which improves the coherence when using various text processing operations. In order to build the vocabulary, we first text-normalized our dictionaries as follows:

```
Method: Normalize
Inputs:
  Word: String
  Arabic_dic: Arabic dictionary
Output:
  Normalized_word: The word after normalization
Begin
  If Word is written in Arabic letters then:
     Normalized_word= replace('أ' with 'ا' ,'إ' with 'ا' ,'آ' with 'ا' ,'ة' with 'ه','ى' with 'ي',
                              'ﻵ' with 'ﻻ', and '-', with '') ;        # '-' elongation character
     If (Normalized_word starts with 'ال') and length(Normalized_word) > 3 and
        (Normalized_word in Arabic_ dic) and (remove_al(Normalized_word) in Arabic_dic)
     then:
        Remove 'ال' from Normalized_word;
     End if
  Else: #Word is written in Latin letters
     Normalized_word = Remove from Word all accents and transform it to lower case; #foreign word
     Remove the apostrophe and the determinant  from Normalized_word;
  End if
Return Normalized_word ;
End.
```

In the previous algorithm, the function remove_al(word) takes a word and returns the same word after removing the definite article "ال".

Here is the algorithm to build the vocabulary.

---

[3]http://www.gwicks.net/justwords.htm.

```
Algorithm: Build the vocabulary of words used in the corpus of posts
Inputs:
        Arabic_dic, English_dic, French_dic: Arabic dictionary, English dictionary, and French diction-
ary, respectively.
        word_list: list of the words of all the posts (without repetitions);
Outputs:
        Arabic_voc, English_voc, French_voc, Dialect_voc_arabic, Dialect_voc_latin:Arabic vocabu-
lary,
        English vocabulary, French vocabulary, dialect using Arabic letters, and dialect written using
Latin letters, respectively.
Begin
        For each word in word_list do:
                Normalized_word= normalize(word) ;
                If Normalized_word is written using Latin letters then :
                        If is_in(Normalized_word, French_dic) then :
                                Write(French_voc, Normalized_word) ;
                        Else If is_in(Normalized_word, English_dic) then :
                                Write(English_voc, Normalized_word) ;
                        Else :
                                Write(Dialect_voc_latin, Normalized_word) ;
                Else :
                        If is_in(Normalized_word, Arabic_dic) then :
                                Write(Arabic_voc, Normalized_word) ;
                        Else :
                                Write(Dialect_voc_arabic, Normalized_word) ;
                End if
        End for
End.
```

The method is_in(Word, Dictionary) replaces any sequence of more than 2 occur-rences of the same letter by only two, before looking up the word in the dictionary. If it does not find it, it checks with only one occurrence of the repeated letter. It returns a boolean value Belongs_to as True or False depending on the result.

Algerian Facebook users mix Arabic, Algerian (informal Arabic), French, and English (Fig. 6). It turns out that some Algerian dialect words when transliterated to Latin letters may produce words that exist in English. For instance, the word "وين"(meaning, where) is sometimes written in Latin characters as "wine". To avoid such ambiguities, we have manually tagged the English words that may be ambiguous

Fig. 6 Language content ratios in Algerian Dialect posts on Facebook

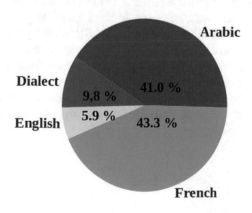

in this sense and removed them from the English vocabulary. This is to give a priority for these words to be first interpreted as belonging to the Algerian dialect. The result of our analysis is that among the 2442 English words we found in the corpus, we were able to detect 98 words that could be found as transliterations of Algerian dialect words.

We also noticed that the dialectal vocabulary contains noise (non-dialect words). In fact, (523) French and (800) Arabic words were not detected by the dictionaries. In order to solve this problem, we proceeded as follows: we manually labeled the two dialect vocabulary files (dialect file written in Latin letters and dialect file written in Arabic letters) so as to assign a class (*dialect*, *Arabic*, *French* or *other*) to each word and then delete any misplaced one and reinsert it into the appropriate file. The 'other' label represents words that do not belong to any of the vocabularies.

4.1.4 Data Normalization

Data normalization is an important step in machine learning in general, and in artificial neural networks more specifically. Since different attributes may take values in ranges of very different scales, this may affect the extent to which each attribute will contribute in the computations of the different neurons. Data normalization, which is different from text normalization that we explained in the previous sub-section, brings all the attribute values back into the same interval of values. To do this, we proceed as follows: for continuous numeric variables whose maximum and minimum values are known, the min-max formula is used, as is the case with dominant color values. The other attributes have been discretized so as to categorize and encode them afterwards.

4.1.5 Dimensionality Reduction

Some attributes may turn out to be correlated which causes data redundancy which unnecessarily increases the dimensions (attributes) of the problem. To avoid this, after the extraction of the attributes, we found by expertise or by correlation study that some attributes are redundant and some others are useless not adding any significant information towards the building of our model. Since, Facebook does not display the actual size of the image, we have removed the dimensions of the images from the dataset. In the same way, we have removed the attributes involved in the calculation of the engagement rate, such as the number of "likes", the number of comments, the number of shares, etc. We have also removed the year of publication to be able to make predictions for future posts, i.e. to be published in years not covered by the training set.

4.1.6 Data Discretization

Data discretization [27] consists in transforming attributes with continuous values into ones that can take a discrete number of values, one for each of an interval of values of the original ones. The aim is to reduce the search space of an algorithm, such as a machine learning algorithm. The techniques we have used to build our models do not require that all attributes be discretized. Nevertheless, discretization can be used to improve the quality of the learning. We have discretized the duration of the video and the number of images, because we argue that this is intuitively useful. Likewise, the engagement rate can also be discretized and transformed into a value belonging to one of several subintervals. For videos for instance, we have divided their durations into 3 sub-intervals: short videos with a duration of less than 1 min, average-length videos with a duration of more than a minute and less than 5 min, and long videos for the rest. These sub-intervals were fixed by expert digital marketers from Sense Conseil.

At this point, we had constructed a dataset with 21,859 Facebook posts collected from 23 pages of Algerian brands, 15% of the collected dataset having been left for testing.

4.2 Building the Learning Model

Once the data is collected and pre-processed, the learning phase takes place. One should recall that the characteristics of a post can be divided into three: the textual part of the post, the image/video part, and the "rest of the characteristics" which can have an impact on the engagement rate just like the textual and visual characteristics. An example of this third category of features/characteristics is the publication time which, according to all expert digital marketers, is an important feature to take into account. These three categories of characteristics are handled in the two proposed learning models as shown in Figs. 7 and 8.

The learning that gets performed on the data is a supervised learning that predicts the engagement rate of a new publication knowing the characteristics of previous publications and the values of their engagement rates. The problem of predicting the engagement rate is a regression problem. Nevertheless, it can be transformed into a classification problem by applying the discretization of the engagement rate on its domain of definition as explained in the previous section.

To predict the engagement rate of a Facebook post, we propose two deep learning models: a multilayer perceptron (MLP) [29] architecture (Fig. 7) and a hybrid neural network architecture (Fig. 8). Regarding the first type of models, we have applied different configurations. Each configuration is characterized by the number of layers used (between 2 and 4 layers), the number of neurons in each layer, the activation function used on the hidden and the output layers (Tanh, Sigmoid, Relu or Softplus) and the number of epochs (iterations).

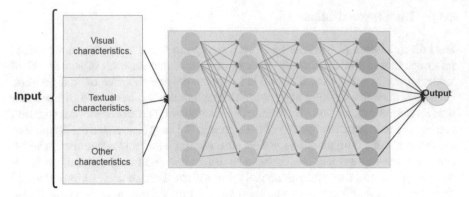

Fig. 7 MLP architecture for engagement rate prediction

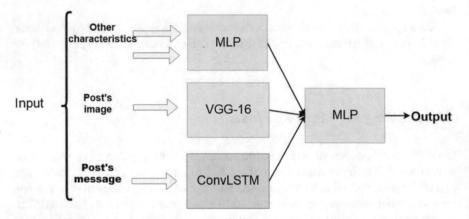

Fig. 8 Hybrid architecture for engagement rate prediction

As to the second type of models, we used, on the one hand, the VGG-16 network, a very deep Convolutional network [30], without its classification layers (its last three layers) to extract the visual characteristics of the image. During training, the hyper parameters of these layers were fixed by those of the VGG-16. On the other hand, we implemented a Conv-LSTM network for the extraction of the textual characteristics of the message. This network consists of a convolution layer with 32 size filters (convolutions) with 'ReLu' as an activation function and a 'Pooling' layer followed by an LSTM layer. The outputs of these two networks (VGG-16 and LSTM) are concatenated with the rest of the parameters and fed into a fully connected MLP whose layers are in turn connected to the output layer.

We note that a conv-LSTM network [31] is a combination of a convolution neural network (CNN) with a Long Short-Term Memory neural network [29]. As such, the strength of CNN in automatically extracting features, e.g. from an image, is combined with the strength of Recurrent Neural Networks such as LSTM to handle time-dependent data. We have chosen to use C-LSTM to anticipate an extension of this work where videos will be processed more thoroughly, and C-LSTM can thus be a very natural choice (Images and time series) [32, 33].

In order to evaluate our models, we opted for Mean Squared Error (MSE) and Mean Absolute Error (MAE) as shown in the results section. Although people may decide to use one of these metrics or the other, it turns out that, based on the datasets on which they get used, the performance may stagnate with one of these metrics while it may continue to increase/decrease with the other [28]. Thus, since a model can give a good Mean Squared Error (MSE) and a poor Mean Absolute Error (MAE) or the opposite, we have preferred to observe both as shown in the results section. Furthermore, both metrics are important when dealing with the engagement rate.

5 Results

5.1 Multilayer Network Models

Table 1 shows the results obtained for the different MLP architectures and different activation functions on the hidden layers. In these tests we set the maximum epochs hyper parameter to 100 and the activation function of the output layer to 'Softplus'. The architecture consisted of 3–4 hidden layers with variable numbers of neurons on the different layers. We have selected a few of these architectures and we report them in the tables below under the column labelled "Architecture".

For the different architectures, the training MSE values as well as the test MSE values are between 10^{-4} and 10^{-6}. These values are very low. Similarly, the MAE values are also low, less than 10^{-2}. Among the architectures mentioned above, we

Table 1 Results obtained for MLP regression models

#	Architecture	Activation Function	MAE	MAE on Test Data	MSE	MSE on Test Data	Training duration
1	100-100-100	Relu	0.0025045	0.0034809	0.0002157	0.0001012	0:29:12.993242
2	100-100-100	Sigmoid	0.0034941	0.0030849	0.0004959	5.33e-05	0:29:19.521814
3	50-50-50	Sigmoid	0.0037052	0.0032972	0.0004959	5.35e-05	0:18:11.418373
4	200-100-50	Sigmoid	0.0034756	0.0030663	0.000496	5.33e-05	0:53:02.034471
5	100-100-100-100	Sigmoid	0.0037261	0.0033181	0.000496	5.36e-05	0:29:28.052561
6	200-100-50	Relu	0.0024584	0.0020503	0.000502	5.74e-05	0:54:32.626196
7	100-100-100-100	Relu	0.0024584	0.0020503	0.000502	5.74e-05	0:29:46.094556
8	**50-50-50**	**Relu**	**0.0017462**	**0.0028421**	**1.84e-05**	**6.79e-05**	**0:18:09.212806**
9	50-50-50	Softplus	0.0017561	0.0022249	2.52e-05	5.94e-05	0:18:09.872247
10	100-100-100-100	Softplus	0.0018338	0.0022022	3.35e-05	5.61e-05	0:30:12.760238
11	200-100-50	Softplus	0.0018187	0.0020095	3.67e-05	5.35e-05	0:54:31.488187
12	100-100-100	Softplus	0.0018979	0.002319	4.32e-05	5.68e-05	0:30:27.440913

Table 2 Results obtained for hybrid regression models

#	Architecture	Activation function	MSE	MSE on Test Data	MAE	MAE on Test Data	Training duration
1	**50-50**	**Relu**	**0.0008165**	**0.0000904**	**0.0025898**	**0.0022125**	**1:48:24.648884**
2	50-50	Softplus	0.0008165	0.0000904	0.0025903	0.0022131	1:33:18.968175
3	50-50-50	Relu	0.0008165	0.0000904	0.0025958	0.0022189	1:28:26.516501
4	50-50-50	Softplus	0.0008165	0.0000904	0.0025968	0.0022199	1:28:16.205537
5	100-100-100	Relu	0.0008165	0.0000904	0.0025904	0.0022133	1:32:43.230382
6	100-100-100	Softplus	0.0008165	0.0000904	0.0025658	0.0021843	1:32:13.979258
7	200-200-100	Relu	0.0008165	0.0000904	0.0025936	0.0022166	1:41:20.025334
8	200-200-100	Softplus	0.0008165	0.0000904	0.0025923	0.0022150	1:45:49.206058
9	100-100-100-100	Relu	0.0008165	0.0000904	0.0025654	0.0021845	1:45:09.378723
10	100-100-100-100	Softplus	0.0008165	0.0000904	0.0025898	0.0022125	1:37:07.239293

selected the best architecture that gave the lowest value in terms of MSE; it is the one with 3 hidden layers of 50 neurons each.

5.2 Hybrid Neural Network Models

Table 2 shows the results obtained for the different regression models and different activation functions on the hidden layers. We set the number of epochs to 50 and the output activation function to 'Softplus'. For the LSTM layer used for the extraction of textual features, we set the number of neurons to 100.

For the different architectures tested, the MSE values for the learning and testing sets vary between 10^{-4} and 10^{-5}. We have selected the first architecture that has the smallest MAE value.

We also plotted the local regression graphs to determine the dependence between the actual engagement rate values and the predicted values. There is a strong linearity relation between the values of the actual engagement rate and the values of the predicted engagement rate for the multilayer model (Fig. 9) in comparison with the hybrid model. The MSE value of the multilayer network is smaller than that of the hybrid network (Fig. 10). Hence our choice goes for the multilayer model.

5.3 Validation

The validation of the MLP model was performed according to two metrics. The first metric calculates the mean distance between the actual engagement rate values and the predicted values for the entire data set (MAE). This metric is also used to validate the behaviour of the model for each page. The second metric assigns a reward (+1) to the model each time the distance between the actual value of the engagement rate and

Fig. 9 Local regression between the actual engagement rate and the predicted engagement rate for the multilayer model

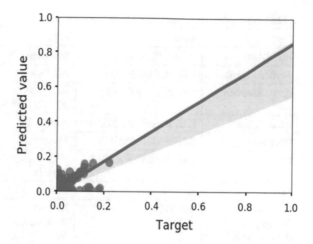

Fig. 10 Local regression between the actual engagement rate and the predicted engagement rate for the hybrid model

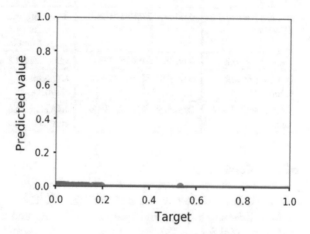

the predicted value is below a certain threshold. This threshold is set for each page. It represents the mean dispersion or absolute deviation values of the engagement rate of a given page. This metric counts the number of tuples whose absolute difference between the value of the engagement rate and the predicted value does not exceed the threshold. Thus, a reward is given to the model if the predicted value is not more dispersed with respect to an average engagement rate than the average tuples.

5.3.1 Metric 1

The mean absolute distance of engagement rates for the entire dataset is 0.0023 (Table 3). Indeed, this distance varies from one page to another. It is very low for certain pages like 'NESCAFÉ', 'ActiviaAlgérie' and 'Amor Benamor' and a little larger for some others like 'Crox Max Chips' and 'Caarama Assurance'.

Table 3 The mean absolute distance of engagement rates for each page

Page	MAE	Page	MAE
NESCAFÉ	0.000205351611341511	Margarine Sol	0.00381443705176773
Activia Algérie	0.000427833407111417	Blédima	0.00409541275951887
Amor Benamor	0.00060543876076198	ALLIANCE ASSURANCES	0.00442537935072357
Afia Algérie	0.000838153822503389	AMANA	0.0049819476616862
Ooredoo Algérie	0.00085011503348794	Macir Vie	0.00500825790292531
Frico	0.00107085709789392	AXA Assurances Algérie	0.0067344885328842
Safina	0.00163100777170885	Margarine Many	0.00711710060155143
CAAT - Compagnie Algérienne des Assurances	0.00179430074193377	La Cash Assurances	0.00725734832472531
Société Nationale D'assurance	0.00269177699171886	Trust Algeria Assurances & Réassurance	0.00910071782887693
Cevital Culinaire	0.00282764962597768	Crox Max Chips	0.0184497466115917
Recettes Nestlé Gloria	0.00371234157095767	Caarama Assurance	0.020034321470615
La Vache Qui Rit Chef	0.00376487706923013	**MAE on the dataset**	**0.0023**

5.3.2 Metric 2

The total score of the data set is 73% (Table 4). This score varies from one page to another. It is very large for most pages like 'Caarama Assurance', 'Crox Max Chips' and very small for the 'Nescafe' page. From the results obtained, the model can be exploited for the prediction of the engagement rate and has actually been validated by Digital Marketing experts.

6 Discussion

Although the results are very encouraging, the data set has some constraints. Thus, we could only use the public data to have the characteristics of the publication and for the calculation of engagement rates. This is why we have recorded engagement rate values that are not well distributed and some are greater than 1, while they should be between 0 and 1. Moreover, as we did not have the number of subscribers at the time of publication, we calculated the engagement rate using the number of subscribers at the time of download. On the other hand, due to the non-use of the 'Page' token, we were unable to retrieve the value of the 'promotion status'

Table 4 The value of the score for each page

Page	Score (%)	Page	Score (%)
Caarama Assurance	97.83	Amor Benamor	80.96
Crox Max Chips	89.4	Ooredoo Algérie	80.68
Trust Algeria Assurances & Réassurance	88.37	La Cash Assurances	80
ALLIANCE ASSURANCES	86.83	Afia Algérie	77.97
Recettes Nestlé Gloria	86.03	Société Nationale D'assurance	77.51
AXA Assurances Algérie	84.9	Blédima	75.47
Margarine Many	82.52	CAAT - Compagnie Algérienne des Assurances	72.9
Macir Vie	82.26	Frico	69.11
AMANA	81.78	Safina	68.34
Cevital Culinaire	81.4	Activia Algérie	60.85
Margarine Sol	81.16	NESCAFÉ	6.27
La Vache Qui Rit Chef	81.05	**Score on the dataset**	**73**

attribute which represents important information. Indeed, sponsored publications are more likely to have a high engagement rate as Facebook increases their visibility by targeting other non-subscribed users who might be interested in the page content. Furthermore, the distribution of publications across all pages is not uniform. These constraints affect the learning; this is why we believe that the very encouraging results we reached can further be improved if the required information is made available.

7 Conclusion

We have presented in this paper a system based on deep learning which allows the prediction of the engagement rate of an advertising content on a social network (Facebook posts). In order to design our prediction system, we first collected posts with their different characteristics from 23 different Facebook pages of commercial brands in Algeria. The posts have been analyzed and pre-processed. Then, we implemented and evaluated several deep learning architectures with different configurations. Finally, among the various models proposed, we selected and validated the best model. Since the size of the current dataset is modest, 21,720 posts from 23 brand pages after pre-processing, we hope that with more data, the results obtained will be further improved especially with the hybrid models since they use the adequate neural network to extract the features (a Conv-LSTM for the textual characteristics, VGG-16 for the visual characteristics and MLP for the other characteristics). That is why we suggest to extend the dataset so that it includes the maximum of pages from the same context (in our case these are pages of commercial brands in Algeria) to collect the maximum of data and thus to improve the different results. Once the dataset is large, it will be interesting to discretize the values of the engagement rate to transform this regression problem into a classification problem and compare the results of the two approaches.

Acknowledgements We express our thanks to the staff of Sense Conseil, especially Mrs. Loubna Lahmici, who were available to answer various questions related to social network marketing.

References

1. We Are Social. *Global Digital Population as of July 2018 (in Millions)*. (July 2018): Statista. https://www.statista.com/statistics/617136/digital-population-worldwide/. (Consulted on 29/08/2018)
2. We Are Social. Kepios; SimilarWeb; TechCrunch; Apptopia; Fortune. *Most Popular Social Networks Worldwide as of July 2018, Ranked by Number of Active Users (in Millions)*. (July 2018): Statista. https://www.statista.com/statistics/272014/global-social-networks-ranked-by-number-of-users/. (Consulted on 29/08/2018)
3. O. Russakovsky, J. Deng, H. Su, J. Krause, S. Satheesh, S. Ma, Z. Huang, A. Karpathy, A. Khosla, M. Bernstein, A.C. Berg, L. Fei-Fei, *ImageNet Large Scale Visual Recognition*
4. A. Soumeur, M. Mokdadi, A. Guessoum, A. Daoud, Sentiment analysis of users on social networks: overcoming the challenge of the loose usages of the Algerian Dialect, in *The 4th International Conference on Arabic Computational Linguistics (ACLing)*, Dubai, 17–19 November, Procedia Computer Science 142 (Elsevier, 2018), pp. 26–37
5. K. Meftouh, N. Bouchemal, K. Smaili, A study of a non-resourced language: an Algerian dialect, in *Proceedings of the Third International Workshop on Spoken Language Technologies for Under-Resourced Languages (SLTU)*, Cape Town, South Africa (2012)
6. H. Saadane, N. Habash, A conventional orthography for Algerian Arabic, in *Proceedings of the Second Workshop on Arabic Natural Language Processing*, Beijing, China, July 26–31 (2015), pp. 69–79
7. Y.M. Kalman, D. Gergle, Letter repetitions in computer-mediated communication: a unique link between spoken and online language, in *Computers in Human Behavior*, (2014, May), pp 187–193
8. Falcon. https://www.falcon.io/. (Consulted on 16/08/2018)
9. Adobe Social. https://www.adobe.com/la/marketing-cloud/social.html. (Consulted on 16/08/2018)
10. S. Wu, M.A. Rizoiu, L. Xie, Beyond views: measuring and predicting engagement in online videos, in *Twelfth International AAAI Conference on Web and Social Media* (2018, June)
11. Z. Chen, A. Hristov, D. Yi, *A Deep Learning Analytic Suite for Maximizing Twitter Impact* (2016, May)
12. Y. Hu, S. Farnham, K. Talamadupula, Predicting user engagement on twitter with real-world events, in *International AAAI Conference on Web and Social Media*, North America (2015, Apr). Available at: https://www.aaai.org/ocs/index.php/ICWSM/ICWSM15/paper/view/10670. Date accessed: 05 Aug 2018
13. B.P. Chamberlain, A. Cardoso, C.H.B. Liu, R. Pagliari, M.P. Deisenroth, *Customer Lifetime Value Prediction Using Embeddings*. arXiv:1703.02596v3 [cs.LG] 6 Jul 2017
14. F. Gelli, T. Uricchio, M. Bertini, A. Del Bimbo, S.F. Chang, Image popularity prediction in social media using sentiment and context features, in *Proceedings of the 23rd ACM International Conference on Multimedia* (ACM, 2015, October), pp. 907–910

15. C. Wei-Fan, C. Yi-Pei, K. Lun-Wei, *How to Get Endorsements? Predicting Facebook Likes Using Post Content and User Engagement* (2017), pp. 190–202. https://doi.org/10.1007/978-3-319-58484-3_15
16. M.X. Hoang, X.H. Dang, X. Wu, Z. Yan, A.K. Singh, GPOP: scalable group-level popularity prediction for online content in social networks, in *Proceedings of the 26th International Conference on World Wide Web*. International World Wide Web Conferences Steering Committee (2017, April), pp. 725–733
17. D.M. Blei, A.Y. Ng, M.I. Jordan, Latent dirichlet allocation. J. Mach. Learn. Res. **3**, 993–1022 (2003)
18. T. Chen, D. Borth, T. Darrell, S.F. Chang, Deepsentibank: Visual Sentiment Concept Classification with Deep Convolutional Neural Networks (2014). arXiv preprint arXiv:1410.8586
19. A. Khosla, A. Das Sarma, R. Hamid, What makes an image popular? in *Proceedings of the 23rd International Conference on World Wide Web* (ACM, 2014, April), pp. 867–876
20. L. Sliburyte, I. Skeryte, What we know about consumers' color perception. Proc. Soc. Behav. Sci. **156**, 468–472 (2014)
21. A.J. Elliot, Color and psychological functioning: a review of theoretical and empirical work. Front. Psychol. **6**, 368 (2015). https://doi.org/10.3389/fpsyg.2015.00368
22. N. Singh, S.K. Srivastava, Impact of colors on the psychology of marketing—a comprehensive overview. Manage. Labour Stud. **36**(2), 199–209 (2011)
23. E.W. Forgy, Cluster analysis of multivariate data: efficiency versus interpretability of classifications. Biometrics **21**(3), 768–769 (1965). JSTOR 2528559
24. P.S. Heckbert, *Color Image Quantization for Frame Buffer Display*. ACM SIGGRAPH'82 Proceedings (1982)
25. T. Uricchio, L. Ballan, L. Seidenari, A. Del Bimbo, Automatic image annotation via label transfer in the semantic space. Pattern Recognition 71 (2016). https://doi.org/10.1016/j.patcog.2017.05.019
26. M. Attia, *Dictionnaire arabe*. https://sourceforge.net/projects/arabic-wordlist/?source=navbar. (Consulted on 16/05/2018)
27. P.N. Tan, M. Steinbach, V. Kumar, *Introduction to Data Mining*, Addison-Wesley (2006)
28. JJ. *MAE and RMSE—Which Metric is Better?* (2016). https://medium.com/human-in-a-machine-world/mae-and-rmse-which-metric-is-better-e60ac3bde13d (Consulted on 28/07/2019)
29. I. Goodfellow, Y. Bengio, A.C. Courville, *Deep Learning* (The MIT Press, Cambridge, MA, 2015)
30. K. Simonyan, A. Zisserman, *Very Deep Convolutional Networks for Large-Scale Image Recognition* (2014). https://arxiv.org/abs/1409.1556
31. X. Shi, Z. Chen, H. Wang, D. Yeung, W. Wong, W. Woo, Convolutional LSTM network: a machine learning approach for precipitation nowcasting, in *NIPS* (2015) pp. 802–810
32. J.R. Medel, A. Savakis, *Anomaly Detection in Video Using Predictive Convolutional Long Short-Term Memory Networks* https://arxiv.org/ftp/arxiv/papers/1612/1612.00390.pdf
33. C. Zhou, C. Sun, Z. Liu, F.C.M. Lau, *A C-LSTM Neural Network for Text Classification* (2015). https://arxiv.org/abs/1511.08630

Predicting Quranic Audio Clips Reciters Using Classical Machine Learning Algorithms: A Comparative Study

Ashraf Elnagar and Mohammed Lataifeh

Abstract This paper introduces a comparative analysis for a supervised classification system of Quranic audio clips of several reciters. Other than identifying the reciter or the closest reciter to an input audio clip, the study objective is to evaluate and compare different classifiers performing the stated recognition. With the widespread of multimedia capable devices with accessible media streams, several reciters became more popular than others for their distinct reciting style. It is quite common to find people who recite Quran in mimicry tone for popular reciters. Towards the achievement of a practical classifier system, a representative dataset of audio clips were constructed for seven popular reciters from Saudi Arabia. Key features were extracted from the audio clips, and different perceptual features such as pitch and tempo based features, short time energy were chosen. A combination of perceptual features were also completed in order to achieve better classification. The dataset was split into training and testing sets (80% and 20%, respectively). The classifier is implemented using several classifiers (SVM, SVM-Linear SVM-RBF, Logistic Regression, Decision Tree, Random Forest, Ensemble AdaBoost, and eXtreme Gradient Boosting. A cross comparative results for all acoustic features and top six subset are discussed for the selected classifiers, followed by fine-tuned parameters from classifiers defaults to optimize results. Finally we conclude with the results that suggest high accuracy performance for the selected classifiers averaging above 90% and an outstanding performance for XGBoosting reaching an accuracy rate above 93%.

Keywords Audio clips · Arabic language · The Quran · Speaker recognition · Classification · Machine learning

A. Elnagar · M. Lataifeh (✉)
University of Sharjah, PO Box 27272, Sharjah, United Arab Emirates
e-mail: mlataifeh@sharjah.ac.ae

A. Elnagar
e-mail: ashraf@sharjah.ac.ae

© Springer Nature Switzerland AG 2020
M. Abd Elaziz et al. (eds.), *Recent Advances in NLP: The Case of Arabic Language*, Studies in Computational Intelligence 874,
https://doi.org/10.1007/978-3-030-34614-0_10

187

1 Introduction

While it is very easy for human to identify a particular speaker by listening to an audio segment, even a non-linguistic queue such as a laugh [1], the automatic identification and classifications of audio clips continues to pose an array of challenges for researchers in related domains. From basic speech pathology, age or gender identification, and forensic studies; an automatic approach for speaker recognition is of great value for a range of applications.

Distinctive features from audio signals that identify a particular speaker are calculated in the feature extraction. The features include several properties such as mean frequency, standard deviation, median frequency, first quantile, third quantile, interquartile range, skewness, kurtosis, spectral entropy, spectral flatness, mode frequency, centroid, peak frequency with highest energy, average of fundamental frequency measured across acoustic signal, minimum and maximum frequency measured across acoustic signal, etc.

Within the context of this work, several reciters are identified through their distinct style, then a classifier is implemented to predict the reciter that is closest to a given Quranic audio clip based on the extracted features. The implementation is carried out with several known classifiers, and the output is later compared across these classifiers according to the overall prediction performance. An automatic reciter's voice identification system extracts audio features mentioned above from the input audio clips, which will be fed into a supervised classifier during the training phase. The classifier is expected to identify the key properties for each of the reciters so that it can predict or identify the closest reciter of a given audio recitation regardless of the verse being recited. To minimize the feature space, the classifier shall utilize a small set of key representative acoustic features.

We selected all reciters from Saudi Arabia. As they adopt similar recitation styles, the inter-variability and consequently features extracted are very close, making the problem addressed in this work more challenging.

We propose a multi-class classifier, also known as ensemble learning. The objective is to assign audio-recitation clip to a specific class out of seven classes representing the seven selected reciters. It is not our intent to analyze the content of the clip, rather a matchup process. In our experiments, we use a dataset which we have constructed with plans for further expansions into a unique recitations corpus. We adopt a supervised approach to classify the audio clips. Several classifiers were used, including Support Vector Machine (SVM), SVM-Linear, Radial Basis Function (SVM-RBF), Logistic Regression (LR), Decision Tree (DT), Random Forest (RF), ensemble Adaptive Boosting (AdaBoost), and Gradient Boosting (XGBoost). We evaluate the performance of the classifier using several kernels for these classifier and other hyperparameters.

It is worth noting that the availability of Arabic audio-based datasets is scarce, and the need for large datasets is a basic requirement for developing reliable classifier systems. Moreover, it will allow researchers to compare their systems on the same benchmark. It is our intention to make this dataset available on Github for researchers to use and pursue further development.

The remaining of the paper is organized as follows: related research is presented in Sect. 2. Section 3 describes the dataset and features used. The selected classifiers are briefly described in Sect. 4. Experimental results are presented in Sect. 5. Finally, we conclude the work in Sect. 6.

2 Literature Review

Some of the early work in audio classifications started initially with basic speech recognition and gender identifications of the speaker based on several proposed features. While some used frequency domain features [2], others investigated the use of physical and perceptual features [3, 4], Beat and Fourier Transform features [5], and spectral flux along with zero crossing rate and short time energy based features [6]. Furthermore, the use of Mel Frequency Spectral Coefficients is reported in [7] while using neural networks for classification, or in combination with SVM [8]. The work in [9, 10] used a combination of acoustic and prosodic features for the classifications of both age and gender. Several others continue to rely on MPEG-7 based features for a range of events and environmental audio classification [11–13].

An SVM is a binary classifier that makes its decisions by constructing an optimal hyperplane that separates the two classes with the largest margin [14]. A thorough comparison of using SVM classifiers with other classifiers such as Linear, Quadratic, Fisher Linear Discriminant, and Nearest-Neighbor is carried out in [15]. It has been shown that SVM is superior to the other selected classifiers. Hence, we adopt using SVM in our work. In fact, SVM is a very popular classifier with high accuracy rate that has been used in other applications on the Arabic language such as [16–18].

While the majority of the above work was carried out on English audio clips, fewer attempts were implemented for other languages such as Hindi [19] and Arabic language using different features [20–24]. In addition, a recent work on further dialectical level of Arabic Emirati-accented speech samples for speaker identification is reported in [25]. As for the precise context of Quran recitations, it has been noted that it was one of the least studied aspects of Islamic culture. While Muslim scholars were concerned mainly with the linguistic aspects and correct pronunciation of the text "Tajweed", others were only concerned with the analysis of the text [26].

It is noteworthy to mention that ten different known Quran recitations or reading methods (Qira' at plural for readings, or Qira'ah for single). In each recitation, there are two famous narrations (Rewayah) that have been authenticated and narrated from qualified and elder licensed scientists (Sheikh) to their students [27, 28]. The most popular reading is that of Hafs Bin Suleiman Al-Kufi on the authority of Asim Al-Koufi which is being recited in Arabia, Levant, Iraq, India, Pakistan and Turkey. The differences of these recitations are related to spoken words, phonemes, intonation, pronunciations of the letters, nasalization (gunnah), pharyngealization (tafxim), the pauses (wagf), and prolongation of the syllables (madd); all of which taken as a codification of the sound of the revelation as it was revealed to the Prophet Muhammad, and as he subsequently rehearsed it with the Angel Gabriel [29].

Furthermore, there is also a noticeable difference to be stated between "Tarteel" and "Tajweed", while the former relates to a hymnody recitation "in proper order" and "with no haste", the latter is of distinct melodic aspects that have been thought to overlap with music [29] with even similar therapeutic influence [30], and even offered as non-pharmacological intervention for anxiety in various settings [31]. Within that direction an automated checking engine for Quranic learners was introduced using Mel-frequency [32], and a most recent speech recognition within recitations as in [33], but very little is to be found on a similar work for automatic recitation classifiers in the literature, reassuring the need for the described work here for the a wide research community and applications.

3 Dataset Construction

The dataset is constructed by extracting several audio clips from audio files of Quran recitation for seven popular reciters based on acoustic properties of the voice. The dataset consists of 2134 recorded voice samples, collected equally among reciters (Each has around 300 audio recitation-clips). Each voice sample is restricted to not exceed 10 seconds in length and pre-processed by acoustic analysis (see wave and tuneR packages), with an analyzed frequency range of 0–280 Hz (human vocal range). The clips are saved in 'wav' format, mono 16 bits. Therefore, we produced a balanced dataset of audio clips. We chose the seven Saudi reciters based on their popularity and most listened to by people. Namely, the reciters are:

- Sheikh Abdul Rahman Al Sudais
- Sheikh Abdullah Basfar
- Sheikh Abu Bakr Al Shatri
- Sheikh Ahmed Al Ajmi
- Sheikh Ali Al Huthaify
- Sheikh Mahir Al Muaiqly
- Sheikh Muhammad Ayyub.

Extracting audio features from an audio file is used to measure acoustic parameters on acoustic signals for which the start and end times are provided. Required features are saved and reciters labels are also added. Pre-processing includes data standardization. Data standardization is shifting the distribution of each attribute to have a mean of zero and a standard deviation of one. This is useful to standardize attributes for the model. Standardization of the dataset is a common prerequisite for a variety of machine learning application. The following 21 common acoustic properties of each voice are measured and included within the properties file (CSV): mean frequency (in kHz), standard deviation of frequency, median: median frequency (in kHz), first quantile (in kHz), third quantile (in kHz), interquartile range (in kHz), skewness, kurtosis, spectral entropy, spectral flatness, mode frequency, frequency centroid, peak frequency with highest energy, average of fundamental frequency measured across acoustic signal, minimum fundamental frequency measured across acoustic signal, maximum fundamental frequency measured across acoustic signal, average of dominant frequency measured across acoustic signal, minimum of dominant frequency measured across acoustic signal, maximum of dominant frequency measured across acoustic signal, range of dominant frequency measured across acoustic signal, modulation index calculated as the accumulated absolute difference between adjacent measurements of fundamental frequencies divided by the frequency range, and label indicating reciters. We used `librosa`[1] package, which includes functions to load audio from disk, compute various spectrogram representations, and a variety of commonly used tools for music analysis. It is easily integrated with the scikit-learn package. We used `librosa` mainly for feature extraction, which includes low-level feature extraction, such as chromagrams, pseudo-constant-Q (log-frequency) transforms, Mel spectrogram, MFCC, and tuning estimation.

4 Classifiers

There are several classifiers that can be used. The role of such classifier is to simply map input data to an existing class. To help select a proper classifier, three criteria is taken into account, which are listed in order of importance, the size of the dataset, speed, and ease of use. Scikit-learn provides multiple classifiers; the K-Neighbors classifier (KNC); Support Vector Machine (SVM); Gaussian Process classifier (GPC); Decision-Tree classifier (DTC); Random-Forest classifier (RFC); MLP classifier; Ada-Boost classifier (ABC); Gaussian NB; and Quadratic Discriminant Analysis classifier (QDAC). It was concluded that the best classifier for the collected dataset is the Support Vector Machine.

Training the model started by dividing them to two separate groups. The first group is the training set which has 80% of the dataset; classification algorithms were applied to the training set first. The second group is the testing set which has 20% of the original dataset; the testing set is used later to produce the final accuracy of the program. The next step is to run all classifiers with the default hyperparameters and best hyperparameters too by searching for the best settings using grid search. We select SVM as an example and detail its fine tuning process. SVM is a collection of algorithms that analyze data to classify and model them, and hyperparameters are high-level parameters (such as kernel, gamma, and C) that are given to SVM to produce tuned models that produce more accurate classifications. By giving an SVM a training set, it would produce an SVM model.

We preform cross validation on the testing data to check the accuracy using k-folds = 10 (K-folds means k divisions, where k = 10 means 10 divisions). Next, we compare the results of the k-fold tests by applying cross-validation techniques using different kernels in order to determine the best kernel for our dataset. We also

---

[1] https://librosa.github.io/librosa/.

investigate the best C, gamma, and polynomial degree values for building a suitable classifier. Finally, we test the SVM model with the best determined hyperparameters values.

The overall block diagram is depicted in Fig. 1. This is an abstract view that projects the flow of processes for developing the best trained model for each classifier. First step (pre-processing phase) involves the loading of the audio files and extracting the features using the librosa package (i.e., the MFCC features). The data standardization, as explained above, is also carried out during this phase. Next, the resulting pre-processed dataset is split into training and testing sets. The subsequent phase implements the classifier model. For each audio clip (x), librosa extracts acoustic features $(h(x))$. We perform two implementations. In the first run, we use all acoustic features while, in the second run, we use a subset of important features only. We denote the selection between all features and a subset of features by the operator or ($\bigoplus$). Such features $(h(x))$ are fed to the selected classifier (machine learning model) in order to produce the predicted label $(\hat{y})$ of the reciter, which is tested against the actual label (y). In order to fine-tune the weight parameters $(\hat{w})$ of the machine learning model, we adopt quality metrics such as the Residual Sum of Squares (RSS), which is the sum of the squares of residuals. It is a measure of the discrepancy between the actual data (y) and an estimated model $(\hat{y})$. Among all estimated models, we choose the one that has the least RSS. Another popular metric used for model tuning is the Maximum Likelihood Estimation (MLE), which attempts at estimating the parameters by maximizing a likelihood function. Either metric is used along with a machine learning algorithm such as the gradient descent algorithm (iterative optimization algorithm) for finding the minimum of a function. To find a local minimum of a function using gradient descent. The objective is to determine the set of features' optimized weights $(\hat{w})$ of the minimum.

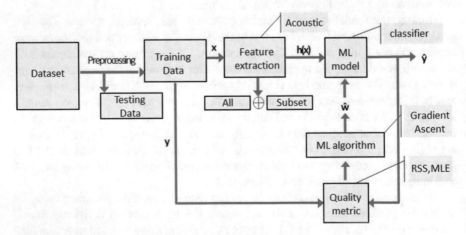

Fig. 1 Abstract flow diagram of the audio classifier system

4.1 Support Vector Machine (SVM), SVM-Linear SVM-RBF

SVM [34], is an elegant and powerful algorithm that is proven to give good results when compared to other classical machine learning algorithms. With a unique advantage in solving small sample and high dimensional pattern recognition problems [35], the SVM algorithms aim at computing a hyperplane that classifies input data points in a n-dimensional features space. Among all possible hyperplanes, the objective is to select the hyperplane that maximizes the separation margins among the predicted classes while allowing for a confident prediction of unseen data points. The hinge loss is used as a loss function to maximize the margin. SVM is a good choice when there is not much information on the input data or the data is not well structured. Although selecting the right kernel function is not straightforward, its use of kernels can treat complex problems. Besides, SVM can scale relatively well to high dimensional data without increasing the risk of over-fitting. However, training an SVM could be a time consuming task when the size of the dataset is large. Cost and gamma are the SVM hyperparameters specifically used to fine-tune the model, which is not an easy task too.

4.2 Logistic Regression (LR)

Logistic Regression [36], is one of the most popular machine learning algorithms. It is easy to implement and produces solid results in a variety of tasks. Therefore every Machine Learning engineer should be familiar with its concepts. This classifier works very well for binary classification as it estimates probabilities of the outcome label and the independent variables (feature space). The Sigmoid function is used to map probabilities into a value between 0 and 1. Hence, it produces a prediction using a threshold classifier. In order to maximize the likelihood that a random data point gets classified correctly, the Maximum Likelihood Estimation employs the Gradient Descent algorithm. The multiclass classification with logistic regression is achieved by either mapping the problem into a binary classification one (one class vs. the rest) or by using the cross-entropy loss function instead.

LR is a widely used technique because it is easy to implement, efficient, less computational resources, highly interpretable, and no fine-tuning is required. However, to obtain good results, it requires some feature engineering. It is not suited for solving non-linear problems. Besides, it is vulnerable to overfitting.

4.3 Decision Tree (DT)

This is another popular supervised learning algorithm [37], which learns decision rules inferred from the training data. As a result, DT algorithm adopts a tree representation in which leaf nodes signify the actual classes to predict and the rest of the nodes represent decisions on features. Order to placing features as root or internal node of the tree is carried out by using a statistical approach such as information gain and gini index. DT rules are intuitive and easy to visualize and interpret. It does not require fine-tuning. However, DT suffers from of overfitting. If information gain is used then the predicted class maybe biased to features with larger categories. Besides, the computations can be complex when there are several classes.

4.4   Random Forest (RF)

Random Forest [38], is an example of ensemble learning model, which aggregates multiple machine learning models in pursuit of higher performance. It is based on multiple decision trees. Unlike DT, which is known for exhibiting high variance and low bias, RF tries to have both low bias and low variance. This will result on predicting classes that are very close to the true ones for cross validation over the dataset. This is achieved by taking a set of high-variance, low-bias decision trees and transform them into a model that has both low variance and low bias. To achieve this goal, RF trains each DT on some part of the training dataset. However, features are selected at random to populate the internal nodes. The resulting DTs shall be not correlated and, therefore, errors are evenly spread and hopefully minimized when the majority voting principle is applied.

RF does not require feature normalization and can be executed in parallel. The problem of overfitting is overcome in RF. It also maintains accuracy even when a large proportion of the data are missing. Of course, RF is complex when compared with DT. Despite parallelism, it requires more computational resources and are less intuitive especially if it involves a large number of decision trees. The prediction process is observed to be time-consuming.

4.5   Adaptive Boosting (AB)

Ada Boost [39], is one of the boosting algorithms that gained significant attention based on its outstanding performance in machine learning competitions. As RF, AB combines an array of low performing classifiers into one model in order to increase performance. As mentioned above, ensemble methods aim at minimizing bias (using boosting) and variance (using bootstrap aggregation). AB differs from RF by the arrangement of base classifiers. While base classifiers are produced in parallel and are homogeneous (i.e., DT) in RF, AB generates them in sequence and are heterogeneous. AB is an iterative ensemble method. AB implements two constraints. First, base classifiers are trained interactively on various training examples. Second, it aims at producing the best fit for the training examples. AB is easy to implement and not prone to overfitting. However, it is slow when compared to other classifiers and highly sensitive to noisy data.

4.6   Gradient Boosting: XGBoost (XG)

XGBoost [40], stands for eXtreme Gradient Boosting. It is an implementation of gradient boosted decision trees in pursuit of higher speed and performance. This machine learning algorithm gained much popularity in the past few years based on its solid performance in machine learning competitions for structured or tabular data. The main objective of this algorithm is efficiency of time and memory resources while training the model. It can easily account for missing data and supports parallelism while constructing its trees. Similar to the above ensemble classifiers, XG implements the gradient boosting decision tree algorithm. Like AdaBoost, XG works by adding models sequentially until no further improvements can be made. However, XG is more sensitive to overfitting especially if the data is noisy. The training process usually takes a bit more time because of the sequential strategy of trees constructions.

5   Experimental Results

As there are no resources available for our application, we constructed a dataset for seven popular selected reciters of the Holy Quran in Saudi Arabia. Usually reciters from the same region, as in our case, follow similar recitation style, which makes it harder to identify the reciter style. Producing the dataset was a time-consuming process, which involved editing, extracting and converting the required audio clips into a set of acoustic features. We used several audio editing software such as WavePad, Audacity Quick Player, and Switch to prepare each audio clip in 'wav' format. The dataset contains 2134 All audio clips of Quranic verses distributed evenly on the selected 7 reciters.

All classifiers were trained on the training dataset of 1707 'wav' audio files and then tested on the testing dataset, which consist of 427 clips. We use 10% of the training dataset for validation purposes while fine tuning each of the classifier's hyperparameters. Space vector representation of acoustic features is used with all supervised classifiers.

In the process of extracting the audio acoustic features, each clip is represented as a vector of 21 features. This step is achieved with warbler package. All vectors of all audio clips are saved to a normal text file for further processing. To evaluate the performance of each classifier, we report the accuracy score, which is simply expressed as the ratio of the number of correctly classified audio clips. We perform polarity as multiclass classification of the popular selected reciters.

We incrementally developed and tested the classifiers listed in Sect. 4 on our training dataset in two main phases. In phase one, we performed classification using all features. During the second phase, we carried out the classification on a subset of features (most important ones).

In this work, a recitation refers to the method of reading of the Quran. There are ten well-known schools of recitations, and each one derives its name from a famous reader of Quran recitation. The most popular recitation is the one of 'Hafs', which is the one adopted by the selected reciters. Recitation should be done according to rules of elocution, pitch, and abruption. During recitation, each melodic passage centers on a single tone level, but the melodic contour and melodic passages are largely shaped by the reading rules, creating passages of different lengths whose temporal expansion is defined through caesuras, hence, creating beautiful melodies of recitations with remarkable and distinct improvisation [41].

As a result, the classification problem becomes harder as all professional reciters strictly apply the rules, but with a melodic distinction of their own.

Reciters may develop different uttered gestures in order to produce the recitation sounds, which is expected to comprise of different concentrations of energy at various frequency locations. Besides, elocution, pitch, abruption, and fricative, and affricate phonemes have energy in many frequency bands. We rely on the produced acoustic features to capture the distinctive features of each reciter and to accurately classify the reciters. As mentioned earlier, pre-processing includes data standardization, which involves shifting the distribution of each attribute to have a mean of zero and a standard deviation of one. This is useful to standardize attributes for the model. Figure 2 depicts the logarithmic scale of all features, used in the classification process, and their importance values to the classification process. Clearly, some features have more impact than other features.

5.1 All Features

Using all acoustic features, we conducted two experiments. In the first one, we tested classifiers using the default settings and reported the average accuracy results of cross validation. In the second one, we fine-tuned the hyperparameters using grid search looking for the best settings for each classifier. The cross validation results for all classifiers are reported in Table 1. Notice that for SVM classifier, we include both linear and rbf kernels.

XGBoost outperformed all classifiers in both experiments. On default settings, LR came as second top while AB scored last. As for the best setting for each classifier, several classifier performed very well producing accuracy scores above 90%. Namely, XG, SVM-linear, LR, and RF. AB scored the least with an accuracy measure of 70%.

As for the best settings for each classifier which were produced by a grid search over the hyperparameters are reported in Table 2.

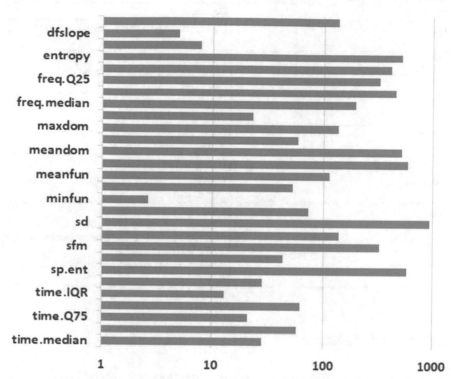

Fig. 2 Importance of acoustic features

Table 1 Performance of classifiers using all features (CV accuracy)

Classifier	Default settings (%)	Best settings (%)
DT	77.50	85.94
RF	85.39	90.51
AB	41.47	70.01
LR	87.20	92.74
SVM-Linear	85.78	93.56
SVM-rbf	83.23	89.87
XG	89.27	94.08

Table 2 The fine-tuned parameters for each classifier on all features

Classifier	Best settings
DT	['max_depth': 10, 'criterion': 'entropy']
RF	['max_depth': 10, 'max_features': 'auto', 'criterion': 'gini']
AB	['n_estimators': 70, 'algorithm': 'SAMME']
LR	['penalty': 'l1', 'solver': 'liblinear', 'C': 1.80]
SVM-Linear	['kernel': 'linear', 'C': 0.9]
SVM-rbf	['gamma': 0.05, 'kernel': 'rbf', 'C': 0.9]
XG	['colsample_bytree': 0.87, 'learning_rate': 0.05, 'max_depth': 5, 'n_estimators': 137,'objective': 'multi:softprob', 'subsample': 0.8]

5.2   Subset of Features

In this experiment, the following acoustic important properties are adopted, namely,

- mean function, "meanfun" (in kHz),
- standard deviation of frequency, "sd",
- first quantile (in kHz), "freq.Q25"
- interquartile range (in kHz), "freq.IQR"
- spectral entropy, "sp.ent"
- spectral flatness measure, "sfm".

The properties of the frequency features (freq.Q25 and freq.IQR) are shown in Fig. 3 above using box plots depicting five statistical measures (minimum, first quartile, median (second quartile), third quartile, and maximum).

Figure 4 displays the relationship among 3 acoustic features, which are mean function, standard deviation, and spectral flatness. The clustering of data points project how each one of these features contribute to the final classification step.

Similarly we show the distribution of data points with respect to spectral entropy and mean frequency in Fig. 5.

Using the above subset of acoustic features, we conducted the same two experiments mentioned earlier. In the first one, we tested classifiers using the default settings and reported the average accuracy results. In the second one, we fine-tuned the hyperparameters using grid search looking for the best settings for each classifier. The cross validation results for all classifiers are reported in Table 3. Accuracy is a

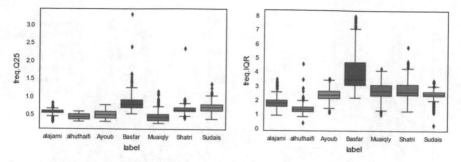

Fig. 3 Acoustic features of two frequency measures

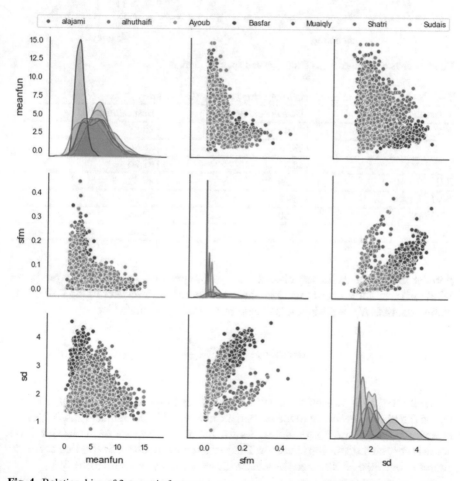

Fig. 4 Relationships of 3 acoustic features

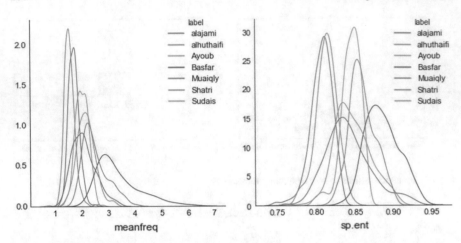

Fig. 5 Acoustic features: mean frequency and spectral entropy

Table 3 Performance of classifiers using 6 top features (CV accuracy)

Classifier	Default settings (%)	Best settings (%)
DT	72.43	80.55
RF	78.35	86.88
AB	48.54	61.80
LR	73.90	80.32
SVM-Linear	81.29	83.36
SVM-rbf	77.50	82.84
XG	80.24	86.00

popular metric for evaluating classifiers. It is the percentage of correctly predicted label (p^i) (for some audio clip i) as the true label t^i with respect to the size of the testing dataset (N). We adopt this metric in all tables. Formally,

$$accuracy = \frac{\sum_{i=1}^{N}(p^i = t^i)}{N}$$

While SVM-Linear outperformed all classifiers in the first experiment, RF produced the highest accuracy in the second experiment. On default settings, XG came as second top while AB scored last. As for the best setting for each classifier, XG score is very close to the top classifier. The rest except AB did well reporting accuracy scores above 80%. AB scored the least with an accuracy measure of 61.8%.

As for the best settings for each classifier which were produced by a grid search over the hyperparameters, using 6 features, are reported in Table 4. Furthermore, Fig. 6 shows a summary of the performance of all classifiers using all and the subset features.

Table 4 The fine-tuned parameters for each classifier on 6 features

Classifier	Best settings
DT	['max_depth': 10, 'criterion': 'entropy']
RF	['max_depth': 12, 'max_features': 'log2', 'criterion': 'entropy']
AB	['n_estimators': 80, 'algorithm': 'SAMME']
LR	['penalty': 'l1', 'solver': 'liblinear', 'C': 4.2]
SVM-Linear	['C': 0.9, 'kernel': 'linear']
SVM-rbf	['C': 0.9, 'gamma': 0.05, 'kernel': 'rbf']
XG	['colsample_bytree': 0.8, 'learning_rate': 0.05, 'max_depth': 6, 'n_estimators': 131, 'objective': 'multi:softprob', 'subsample': 0.8]

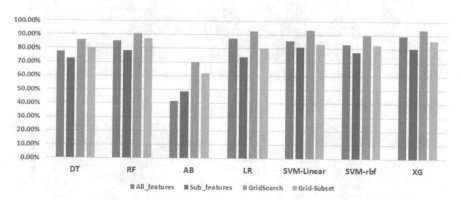

Fig. 6 Comparison of CV accuracy of all classifiers

Using best settings for each classifier whether it is using all features or the subset features, we report the final accuracy results on the testing the 427 unseen audio clips in Table 5. XG came at top with solid performance of 93.44% followed by SVM-Linear. All classifiers did extremely well around 90% except for AB which scored an accuracy of 70.26%.

Figure 7 depicts the number of misclassified audio clips per reciter and per fine-tuned classifier. Notice that the reciter "Shateri" has the most misclassified results while reciter "Huthaifi" has the least. This means that "Huthaifi" features are most distinct among other reciters. Figure 8 displays the normalized confusion matrix for two classifiers as an example. Namely, the SVM-linear classifier (Fig. 8 (left side))

Table 5 Accuracy scores of the fine-tuned classifiers on the testing dataset

Classifier	Accuracy (%)
DT	89.70
RF	89.46
AB	70.26
LR	90.40
SVM	91.57
XG	93.44

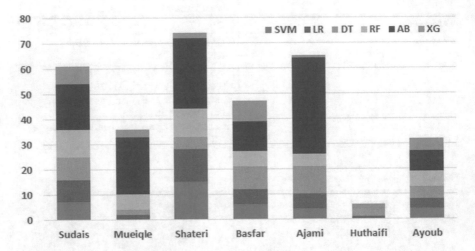

Fig. 7 Misclassifications by reciter

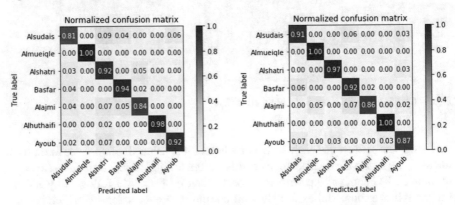

Fig. 8 Normalized confusion matrix of the SVM (left) and XGBoost (right) classifiers

and XGBoost (Fig. 8 (right side)). Confusion matrix (error matrix), is a popular metric in machine learning as a visualization of the performance of a classifier in a form of a table. It displays expected values vs predicted values (true positives, false positives, false negatives, and true negatives).

5.3 Detailed Example: The SVM Classifier Case

During this example, we tested the SVM classifier on two reciters only. We chose SVM because of its solid performance in our experiments above and its wide success in a variety of applications. On polarity classification, SVM using a liner kernel reported an accuracy score of 98.20%. See Table 6 and Fig. 9.

Next, we tested for all reciters. Upon extracting the features and adjusting the hyperparameters, we tested the SVM classifier (default hyperparameter) with a linear kernel, rbf kernel, and a polynomial kernel. The resulting classifier attained an accuracy of 90% using the rbf kernel. Table 6 summarizes system accuracy scores using different SVM implementations during each phase. Figure 10 depicts these findings too.

Table 6 Performance of proposed system during each experiment

No. of reciters	Linear (%)	Polynomial (%)	RBF (%)
2	98.20	78.57	84.76
7	89.70	73.07	89.93

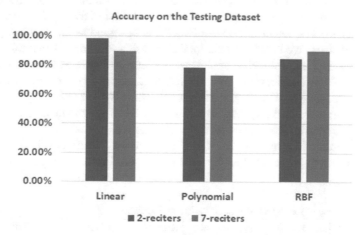

Fig. 9 Performance evaluation of the SVM classifier

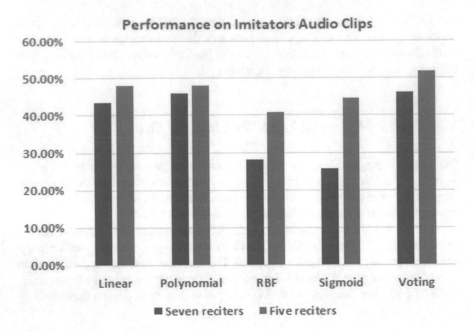

Fig. 10 Performance evaluation on imitators' audio clips

5.3.1 Imitators

In addition, we tested the SVM classifier on audio clips of some people who attempt at imitating popular reciters. We selected few people whose audio clips are available on YouTube. The objective was to predict the closest reciter to the input recitation-clip based on its acoustic features. The results were not as good as the ones obtained on the testing dataset. We tested the imitators' clips using four models instead of SVM. Namely, we used the following kernels to implement the SVM classifier: Linear, Polynomial, RBF, and Sigmoid. We added also a majority vote classifier based on SVM implementations. The final results are shown in Table 7.

The performance of the system is measured by a voting scheme among the listed SVM-models above. The process is to predict the reciter based on counting votes obtained from the four models. For example, if several classifiers vote for the same reciter then it will be selected as the final prediction. In the event of a tie, we use

Table 7 Accuracy of testing SVM models on the imitators' audio clips

No. of reciters	Linear (%)	Polynomial (%)	RBF (%)	Sigmoid (%)	Voting (%)
5	48.13	48.13	40.74	44.44	51.85
7	43.50	46.00	28.21	25.64	46.15

accuracy to resolve the conflict. The accuracy is satisfactory. In another experiment, we dropped two reciters (Sheikh Muhammad Ayyub and Sheikh Ali Al Huthaify) as both were misclassified. Testing results, without these two reciters, increased by almost 7%. Table 7 and Fig. 10 show the accuracy results of the different models on the imitators' audio clips of both experiments. The misclassification is mainly attributed for two reasons. First, the quality of the imitators audio clips is not as good as the ones used for training. We have found that there is a noticeable background noise. Second, some of the imitators poorly imitate the original reciters; far from the recitation style.

5.3.2 SVM Hyperparameters Tuning

We would like to shed light on the results of the classifier on the dataset after training for polarity classification. The dataset is divided into training and testing datasets. The testing dataset consists of 427 audio clips covering the seven reciters. Table 8 shows the confusion matrix for each reciter.

Table 9 shows the accuracy (and mis-classification error) for each reciter. The names of the reciters are displayed, which are in order with the reciter number of Table 8. Notice that reciters "Alajami", "Ayoub", and "Muaiqli" have high misclassification rate compared to "Shatri", "Sudais", and "Alhuthaifi". This means that the features for the first set of reciters are close and overlap. The reciters "Alajami" and "Ayoub" has been mistakenly predicted more than 50% of all misclassification cases. This means both reciters share several features with the other reciters.

We used K-fold cross validation to get the accuracy results. Of course, each fold test would give a different accuracy score as the dataset is split in random manner into testing and training datasets. The dataset is split into 10 equal parts thus covering all the data into training as well into testing sets.

As part of tuning the SVM hyperparameters, we tuned the parameter C, which controls the SVM optimization by specifying the significance of avoiding misclassifying each training example. For large values of C, the optimization will choose a smaller-margin hyperplane. In contrast, a very small value of C will cause the op-

Table 8 Confusion matrix of SVM-rbf kernel

Reciter #	Precision	Recall	F1-score	Support
Alajami	0.86	0.85	0.85	71
Alhuthaifi	0.97	0.96	0.97	72
Ayoub	0.84	0.82	0.83	57
Basfar	0.80	0.90	0.85	59
Muaiqly	0.92	0.85	0.89	55
Shatri	0.98	0.98	0.98	53
Sudais	0.93	0.93	0.93	60

Table 9 Accuracy per reciter

Reciter name	Accuracy (%)
Alajami	84.51
Alhuthaifi	95.83
Ayoub	82.46
Basfar	89.83
Muaiqly	85.45
Shatri	98.11
Sudais	93.33

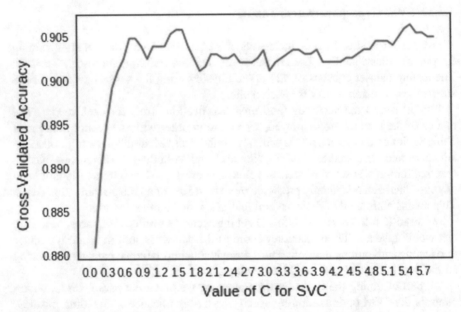

Fig. 11 Optimizing for SVM C value

timizer to look for a larger-margin sepa-rating hyperplane, even if that hyperplane misclassifies more points. The tuning reported that $C = 1.5$ gives the highest accuracy (see Fig. 11).

Another hyperparameter is gamma for kernel as rbf. Technically, the gamma hyperparameter controls the variance. Therefore, a small gamma value would lead a large variance which allow for distant points to be considered similar. Alternatively, a large gamma value means a small variance. Figure 12 displays the result for different values of gamma. Large values lead to poor performance of the model. Best gamma value is 0.01.

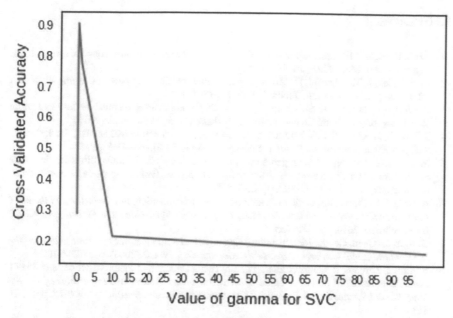

Fig. 12 Optimizing for SVM gamma value

6 Conclusions

In this work, we have developed a multi-class voice classifier system of Quranic audio clips. We constructed the dataset for seven popular Saudi reciters. Representative
acoustic features of the audio clips were extracted and standardized. We implemented
several popular and successful supervised classifiers to identify the reciter of a given
audio Quranic clip. We further fine-tuned all classifiers in pursuit of the best classifier model. We trained the models on all acoustic features as well as top six features.
XGBoost is proven to deliver the best results. The system performance on classifying audio clips exceeded 98% for two reciters using SVM and 93.44% for all seven
reciters. However, testing the performance of the classifier on new clips of some imitators, who mimic, one of the seven reciters varies between 40 and 50% using SVM.
It is expected that the system will not perform as good because the imitators dataset
is noisy and few of the imitators poorly imitate the professional seven reciters. However, the prediction rate is higher than the baseline accuracy which ranges between
14.3% (for the seven reciters) and 20% (for the five reciters).

Acknowledgements We would like to thank Rotana Ismail, Bahja Alattas, and Alia Alfalasi for
initiating the work and constructing the dataset. We extend our thanks to the University of Sharjah
for funding this work under targeted research project no.: 1702141151-P.

References

1. J.H.L. Hansen, T. Hasan, Speaker recognition by machines and humans: a tutorial review. IEEE Signal Process. Mag. **32**(6), 74–99 (2015)
2. M.A. Sadek, I. Md Shariful, H. Md Alamgir, Gender recognition system using speech signal. Int. J. Comput. Sci. Eng. Inf. Technol. **2**(1), 1–9 (2012)
3. F. Alías, J.C. Socoró, X. Sevillano, A review of physical and perceptual feature extraction techniques for speech, music and environmental sounds. Appl. Sci. **6**(5) (2016)
4. J.H. Bach, J. Anemüller, B. Kollmeier, Robust speech detection in real acoustic backgrounds with perceptually motivated features. Speech Commun. 53(5), 690–706 (2011)
5. Y. Yaslan, Z. Cataltepe, Music genre classification using audio features, different classifiers and feature selection methods, in *2006 IEEE 14th Signal Processing and Communications Applications*, vols. 1 and 2 (2006), pp. 535–538
6. A. Ghosal, S. Dutta, Automatic male-female voice discrimination, in *Proceedings of the 2014 International Conference on Issues and Challenges in Intelligent Computing Techniques, ICICT 2014*, February 2014, pp. 731–735
7. H. Harb, L. Chen, Gender identification using a general audio classifier, in *Proceedings of the IEEE International Conference on Multimedia and Expo*, vol. 2 (2003), pp. II733–II736
8. P. Mini, T. Thomas, R. Gopikakumari, Feature vector selection of fusion of MFCC and SMRT coefficients for SVM classifier based speech recognition system, in *Proceedings of the 8th International Symposium on Embedded Computing and System Design, ISED 2018*, pp. 152–157
9. M. Maqsood, A. Habib, T. Nawaz, An efficient mispronunciation detection system using discriminative acoustic phonetic features for Arabic consonants. Int. Arab J. Inf. Technol. **16**(2), 242–250 (2019)
10. H. Meinedo, I. Trancoso, Age and gender classification using fusion of acoustic and prosodic features, in *Interspeech-2010*, January 2010, pp. 2818–2821
11. H. Kim, N. Moreau, T. Sikora, Audio classification based on MPEG-7 spectral basis representations. IEEE Trans. Circuits Syst. Video Technol. **14**(5), 716–725 (2004)
12. C. Okuyucu, M. Sert, A. Yazici, Audio feature and classifier analysis for efficient recognition of environmental sounds, in *Proceedings of the 2013 IEEE International Symposium on Multimedia, ISM 2013* (2013), pp. 125–132
13. J.-C. Wang, J.-F. Wang, K.W. He, C.-S. Hsu, Environmental sound classification using hybrid SVM/KNN classifier and MPEG-7 audio low-level descriptor, in *International Joint Conference on Neural Networks*, 2006, pp. 1731–1735
14. L.T. Christopher, J.C. Burges, A tutorial on support vector machines for pattern recognition. Data Min. Knowl. Discov. **2**, 121–167 (1998)
15. B. Moghaddam, M.H. Yang, Gender classification with support vector machines, in *Proceedings of the 4th IEEE International Conference on Automatic Face & Gesture Recognition, FG 2000*, 2000, pp. 306–311
16. A. Elnagar, Y.S. Khalifa, A. Einea, Hotel Arabic—reviews dataset construction for sentiment analysis applications, in *Studies in Computational Intelligence*, vol. 740, ed. by K. Shaalan, A. Hassanien, F. Tolba (Springer, 2017), pp. 35–52
17. A. Elnagar, A. Einea, Investigation on sentiment analysis of Arabic book reviews, in *IEEE/ACS 13th International Conference of Computer Systems and Applications (AICCSA)*, 2016, pp. 1–7
18. A. Elnagar, O. Einea, Book reviews in Arabic dataset, in *IEEE/ACS 13th International Conference of Computer Systems and Applications (AICCSA)*, 2016, pp. 1–8
19. A. Pahwa, G. Aggarwal, Speech feature extraction for gender recognition. Int. J. Image Graph. Signal Process. **8**(9), 17–25 (2016)
20. S.S. Al-Dahri, Y.H. Al-Jassar, Y.A. Alotaibi, M.M. Alsulaiman, K. Abdullah-Al-Mamun, A word-dependent automatic Arabic speaker identification system, in *IEEE International Symposium on Signal Processing and Information Technology*, 2008, pp. 198–202

21. A. Krobba, M. Debyeche, A. Amrouche, Evaluation of speaker identification system using GSMEFR speech data, in *International Conference on Design & Technology of Integrated Systems in Nanoscale Era*, 2010, pp. 1–5
22. A. Mahmood, M. Alsulaiman, G. Muhammad, Automatic speaker recognition using multi-directional local features (MDLF). Arab. J. Sci. Eng. **39**(5), 3799–3811 (2014)
23. H. Tolba, A high-performance text-independent speaker identification of Arabic speakers using a CHMM-based approach. Alexandria Eng. J. **50**(1), 43–47 (2011)
24. K. Saeed, M.K. Nammous, A speech-and-speaker identification system: feature extraction, description, and classification of speech-signal image. IEEE Trans. Ind. Electron. **54**(2), 887–897 (2007)
25. I. Shahin, A.B. Nassif, M. Bahutair, Emirati-accented speaker identification in each of neutral and shouted talking environments. Int. J. Speech Technol. **21**(2), 265–278 (2018)
26. F.M. Denny, Qur'an recitation: a tradition of oral performance and transmission. Oral Tradit. **4**(1–2), 5–26 (1989)
27. S.A.E. Mohamed, A.S. Hassanin, M. Taher, B. Othman, *Virtual Learning System (Miqra'ah) for Quran Recitations for Sighted and Blind Students* (2014), pp. 195–205
28. B. Yousfi, Holy Qur'an speech recognition system Imaalah checking rule for Warsh recitation, in *Proceedings of the 13th International Colloquium on Signal Processing and its Applications, CSPA 2017*, pp. 258–263. https://doi.org/10.1109/CSPA.2017.8064962
29. K. Nelson, *The Art of Reciting the Qur'an* (University of Texas Press, 1985)
30. A. Abdurrochman, R.D. Wulandari, N. Fatimah, The comparison of classical music, relaxation music and the Qur'anic recital: an AEP study, in *The 2007 Regional Symposium on Biophysics and Medical Physics*, November 2007
31. A. Ghiasi, The effect of listening to holy quran recitation on anxiety: a systematic review. Iran. J. Nurs. Midwifery Res. **23**(6), 411–420 (2018)
32. J.I. Noor, M. Razak, N. Rahman, Automated tajweed checking rules engine for Quranic learning. Multicult. Educ. Technol. J. **7**(4), 275–287 (2013)
33. F. Thirafi, Hybrid HMM-BLSTM-based acoustic modeling for automatic speech recognition on Quran recitation, in *Proceedings of the International Conference on Asian Language Processing, IALP 2018*, pp. 203–208. https://doi.org/10.1109/IALP.2018.8629184
34. J.A.K. Suykens, J. Vandewalle, Least squares support vector machine classifiers. Neural Process. Lett. **9**(3), 293–300 (1999). https://doi.org/10.1023/A:1018628609742
35. L. Sun, Decision tree SVM model with Fisher feature selection for speech emotion recognition. Eurasip J. Audio Speech Music Process. **1**(1) (2019)
36. D.W. Hosmer Jr., S. Lemeshow, R.X. Sturdivant, *Applied Logistic Regression*, 3rd edn. (Wiley, 2013). https://doi.org/10.1002/9781118548387
37. J.R. Quinlan, Induction of decision trees. Mach. Learn. **1**(1), 81–106 (1986). https://doi.org/10.1007/BF00116251
38. A. Liaw, M. Wiener, Classification and regression by Random Forest. R News **2**, 18–22 (2002)
39. Y. Freund, R.E. Schapire, A short introduction to boosting. J. Jpn. Soc. Artif. Intell. **14**(5), 771–780 (1999)
40. J.H. Friedman, Greedy function approximation: a gradient boosting machine. Ann. Stat. **29**, 1189–1232 (2001)
41. T. Al Bakri, M. Mallah, Musical performance of the holy Quran with assistance of the Arabic Maqams. Turk. Online J. Educ. Technol. **2016**, 121–132 (2016)

Printed in the United States
By Bookmasters